Lecture Notes in Mathematics

Edited by A. Dold and B. Eckmann

692

T. Husain
S. M. Khaleelulla

Barrelledness in Topological and Ordered Vector Spaces

Springer-Verlag
Berlin Heidelberg New York 1978

Authors
T. Husain
Mathematics Department
McMaster University
Hamilton, Ontario
Canada, L8S 4K1

S. M. Khaleelulla
Department of Mathematics
Malnad College of
Engineering
Hassan – 573201/India

AMS Subject Classifications (1970): 46 A 07, 46 A 40

ISBN 3-540-09096-7 Springer-Verlag Berlin Heidelberg New York
ISBN 0-387-09096-7 Springer-Verlag New York Heidelberg Berlin

Printing and binding: Beltz Offsetdruck, Hemsbach/Bergstr.
2141/3140-543210

PREFACE

It is well known that in Functional Analysis the notion of barrelled space plays an important role. For example, the open mapping and closed graph theorems and the Banach-Steinhaus theorem which are true for Banach spaces require an additional property on locally convex spaces for their validity and that is barrelledness. A number of generalizations of barrelled spaces have since been studied. The aim of these authors in writing this monograph is to give an account of certain classes of topological vector spaces and ordered topological vector spaces which generalize barrelled spaces and also to see if certain theorems true for Banach space can be carried over to these generalized classes of spaces. To mention a few generalizations of the above-mentioned notion, they include, quasi-barrelled, countably barrelled, σ-barrelled spaces etc. We wish to give an up-to-date listing of such spaces and their usefulness in carrying over certain classical theorems in this monograph.

This book consists of nine chapters, starting with Chapter 1 in which elementary results in topological and ordered topological vector spaces are given. Their proofs are omitted since they are supposed to be known to the reader and are available in any standard book on topological vector spaces or ordered topological vector spaces, e.g. Bourbaki [4], Köthe [31], Schaefer [47], Horváth [11], Peressini [41], Namioka [39] or Wong and Ng [63].

Chapter 2 also consists of known results on classical locally convex topological vector spaces and can easily be found in [4], [12], [31], [47] or [10].

In Chapter 3, we study ultrabarrelled, ultrabornological and quasiultrabarrelled spaces. They were first introduced and studied by W. Robertson [44], and Iyahen [19].

Chapter 4 deals with order-quasibarrelled vector lattices which were first introduced by Wong [61]. A kind of closed graph and Banach-Steinhaus theorems for lattice maps has been proved.

In Chapter 5, the notions of countably barrelled and countably quasibarrelled spaces which were first introduced by the first author [13] have been considered. Among the results proved therein, the Banach-Steinhaus theorem is the most important one. These notions naturally lead to a notion what is called here as H-spaces (H stands for Husain) which was first studied by the second author [26].

The contributions of De Wilde and Houet [7], and Webb [56] regarding σ -barrelled and sequentially barrelled spaces which generalize the countably barrelled spaces are given in Chapter 6. Their application to locally convex spaces with bases are given.

Chapter 7 includes the study of countably ultrabarrelled and countably quasiultrabarrelled spaces and their tensor product. These spaces need not be locally convex and so the application of the duality theorem being unavailable, direct methods are used in proving the analogues of Banach-Steinhaus and isomorphism theorems.

In Chapters 8 and 9, the notions of Chapter 5 and 3 are studied when the vector space is ordered and thus similar results are proved, for order preserving maps, including a closed graph theorem for such maps.

In the study of such spaces, there have remained a number of open problems which are listed at the end of each chapter. The last chapter is followed by a bibliography which is followed by an Index. The numbering of defintions, theorems etc. in each chapter starts afresh. A reference not accompanied by the number of the chapter refers to the chapter in which it occurs.

The authors are grateful to the secretaries, in particular Miss Kay Chornook, of the Department of Mathematics at McMaster University for typing the manuscript.

T. Husain
and
S.M. Khaleelulla

C O N T E N T S

CHAPTER I

PRELIMINARIES

CHAPTER II

SOME CLASSES OF LOCALLY CONVEX SPACES

CHAPTER III

ULTRABARRELLED, ULTRABORNOLOGICAL AND QUASIULTRABARRELLED SPACES

CHAPTER IV

ORDER-QUASIBARRELLED VECTOR LATTICES AND SPACES

CHAPTER V

COUNTABLY BARRELLED AND COUNTABLY QUASIBARRELLED SPACES

CHAPTER VI

σ -BARRELLED AND SEQUENTIALLY BARRELLED SPACES

CHAPTER VII

COUNTABLY ULTRABARRELLED AND COUNTABLY QUASIULTRABARRELLED SPACES

CHAPTER VIII

COUNTABLY ORDER-QUASIBARRELLED VECTOR LATTICES AND SPACES

CHAPTER IX

ORDER-QUASIULTRABARRELLED VECTOR LATTICES AND SPACES

CHAPTER 1
PRELIMINARIES

In order to facilitate the reading of this book, some of the important notions and results about topological vector spaces and ordered topological vector spaces are collected in this chapter. The proofs of these results are omitted, since they are readily available in standard books on topological vector spaces, e.g. Bourbaki [4], Köthe [31], Schaefer [47] and on ordered topological vector spaces, e.g., Peressini [41] and Namioka [39].

§1. Topological vector spaces

Definition 1. A vector space E (over the field $\mathbb{R}$ of real numbers) equipped with a (Hausdorff) topology u is called a topological vector space (in short, (Hausdorff) t.v. space) if the maps

$$(x,y) \rightarrow x + y,$$

and

$$(\lambda,x) \rightarrow \lambda x$$

are continuous in both variables together for x,y ε E and $\lambda \varepsilon \mathbb{R}$. We write $(E,\mathfrak{u})$ to represent a t.v. space E with topology $\mathfrak{u}$. $\mathfrak{u}$ is called a vector or linear topology on E. All topological vector spaces are assumed to be Hausdorff unless otherwise stated.

Remark 1. In order to have consistency throughout, we shall deal with only real vector spaces and Hausdorff t.v. spaces, although most results are valid for complex vector spaces.

Definition 2. Let E be a vector space.

(i) A subset A of E is said to be circled if $\lambda A \subset A$ for every $\lambda \in \mathbb{R}$ with $|\lambda| \leq 1$.

(ii) A subset A of E is said to absorb another subset B of E, if there exists an $\alpha > 0$ such that $B \subset \lambda A$ for all $\lambda \in \mathbb{R}$ with $|\lambda| \geq \alpha$.

(iii) A subset A of E is said to be absorbing if for each $x \in E$ there is an $\alpha > 0$ such that $x \in \lambda A$ for all $\lambda \in \mathbb{R}$ with $|\lambda| \geq \alpha$.

(iv) A subset A of E is convex if for all $x,y \in A$, $\lambda x + (1-\lambda)y \in A$, $0 \leq \lambda \leq 1$.

(v) Let A be a non-empty subset of E. The smallest convex (circled) set containing A is called the convex (circled) hull of A.

Theorem 1. In a t.v. space $(E,\mathfrak{U})$, there exists a neighbourhood basis $\mathcal{B}$ of closed neighbourhoods of the origin 0 such that

(N_1) each $U \in \mathcal{B}$ is circled and absorbing;

(N_2) for each $V \in \mathcal{B}$ there is a $V \in \mathcal{B}$ such that $V + V \subset U$;

(N_3) $\cap\{U: U \in \mathcal{B}\} = \{0\}$.

Conversely, if E is a vector space and $\mathcal{B}$ a filter base [25] satisfying

$(N_1) - (N_3)$, then there exists a unique topology $\mathcal{U}$ on E such that $(E,\mathcal{U})$ is a t.v. space and $\mathcal{B}$ is a neighbourhood basis of 0.

Definition 3. A t.v. space $(E,\mathcal{U})$ is said to be metrizable if there exists a countable neighbourhood basis of 0. These neighbourhoods can be so chosen as to satisfy $(N_1) - (N_3)$ of Theorem 1.

Definition 4. Let E be a vector space. An $\mathfrak{F}$-seminorm on E is a map $f: E \to \mathbb{R}^+$ such that

(i) $f(x+y) \leq f(x) + f(y)$;

(ii) $f(sx) \leq f(x)$, $s \in \mathbb{R}$, $|s| \leq 1$;

(iii) $f(sx) \to 0$, when $s \to 0$ in $\mathbb{R}$ and $x \in E$ is fixed.

Remark 2. In addition, if $f(x) = 0 \iff x = 0$, then f is called an $\mathfrak{F}$-norm.

Proposition 1. A vector space topology on E can always be determined by a family of $\mathfrak{F}$-seminorms. If the topology is metrizable, then it can be determined by a single $\mathfrak{F}$-norm.

Theorem 2. Let (E,u) be a t.v. space.

(a) For each $x_0 \in E$ and each $0 \neq \lambda \in \mathbb{R}$, the map $x \to \lambda x + x_0$ is a homeomorphism of E onto itself.

(b) For any subset A of E and any base $\mathcal{B}$ of the neighbourhood filter of 0, the closure $\bar{A}$ of A is given by

$$\bar{A} = \cap\{A + V;\ V \in \mathcal{B}\}.$$

(c) If A is an open subset of E and B any subset of E, then A + B is open.

(d) If A is a closed subset of E and B is a compact subset of E, then A + B is closed.

(e) If A is a circled or convex subset of E, then its closure $\bar{A}$ is also circled or convex, and the interior int(A) of A is circled if $0 \in \text{int}(A)$.

Proposition 2. (a) Let $\{(E_\alpha, \mathcal{U}_\alpha);\ \alpha \in A\}$ be a family of t.v. spaces. Then $E = \prod_\alpha E_\alpha$ is a t.v. space under the product topology $\mathcal{U} = \prod_\alpha \mathcal{U}_\alpha$.

(b) Let $(E, \mathcal{U})$ be a t.v. space and M a subspace of E.

(i) The closure $\bar{M}$ of M in $(E, \mathcal{U})$ is again a subspace of E.

(ii) E/M is a linear space with the quotient topology such that the maps : $(\dot{x}, \dot{y}) \to \dot{x} + \dot{y}$ and $(\lambda, \dot{x}) \to \lambda\dot{x}$ are continuous, when $\dot{x} = x + M$.

(iii) E/M is Hausdorff iff M is closed in $(E, \mathcal{U})$.

The codimension of a subspace M of a vector space E is the dimension of E/M; N is an algebraic complementary subspace of M if E = M + N is an algebraic direct sum.

Proposition 3. (a) Each one dimensional t.v. space (E,u) is isomorphic with $\mathbb{R}$.

(b) Each t.v. space of finite dimension n is isomorphic with $\mathbb{R}^n$.

(c) Let (E,u) be a t.v. space, M a closed subspace and N a finite dimensional subspace. Then M + N is closed in (E,u).

(d) Let (E,u) be a t.v. space and M a closed subspace of finite codimension. Then $E = M \oplus N$ for every algebraic complementary subspace N of M, where $\oplus$ is the direct sum.

(e) A t.v. space is finite dimensional iff it is locally compact.

Definition 5. Let (E,u) be a t.v. space.

(a) A subset B of (E,u) is called bounded if it is absorbed by every neighbourhood of 0.

(b) A fundamental system of bounded sets in (E,u) is a family $\mathfrak{F}$ of bounded sets such that each bounded set in (E,u) is contained in some member of $\mathfrak{F}$.

(c) A subset B of (E,u) is called precompact if for each neighbourhood V of 0 in (E,u) there exists a finite subset $B_0 \subset B$ such that $B \subset B_0 + V$.

Theorem 3. Let (E,u) be a t.v. space.

(a) If A is bounded (respectively, precompact), so is every subset of A.

(b) If A is bounded (respectively, precompact), so are $\overline{A}$ and λA, for any $\lambda \in \mathbb{R}$.

(c) If A and B are bounded (respectively, precompact), so are $A \cup B$ and $A + B$.

(d) Every precompact set in E is bounded. The circled hull of a bounded set is bounded.

(e) The continuous linear image of a bounded (precompact) set in a t.v. space is bounded (precompact).

<u>Corollary 1</u>. The union and the sum of a finite number of bounded sets is again bounded.

<u>Corollary 2</u>. Every Cauchy sequence in a t.v. space is bounded.

<u>Proposition 4</u>. Let $\{(E_\alpha, u_\alpha);\ \alpha \in A\}$ be a family of t.v. spaces and $(E,u) = \prod_\alpha (E_\alpha, u_\alpha)$. Then a subset B of E is bounded iff $B \subset \prod_\alpha B_\alpha$ where each B_α is bounded in (E_α, u_α).

Let E be a vector space and $\{(E_\alpha, u_\alpha);\ \alpha \in A\}$ be a family of t.v. spaces. Let f_α be a linear map of E_α into E and let $\bigcup_\alpha f_\alpha(E_\alpha)$ span E. The family Φ of all vector topologies on E for which each f_α is continuous is not empty, because it contains the trivial topology. The upper bound u of the members of Φ is also in Φ; u is the finest vector topology on E for which all the f_α are continuous.

<u>Definition 6</u>. u as defined above, is called the *-inductive limit topology on E and (E,u) is the *-inductive limit of $\{[(E_\alpha, u_\alpha):\ f_\alpha];\ \alpha \in A\}$.

<u>Proposition 5</u>. Let (E,u) be the *-inductive limit of $\{[(E_\alpha, u_\alpha):\ f_\alpha];\ \alpha \in A\}$.

(a) A linear map g of (E,u) into any t.v. space F is continuous iff $g \circ f_\alpha$ is a continuous linear map of E_α into F for each $\alpha \in A$.

(b) A set $\mathcal{T}$ of linear maps of (E,u) into any t.v. space F is equicontinuous iff $\mathcal{T} \circ f_\alpha$ is an equicontinuous set of linear maps of E_α into F for each $\alpha \in A$.

The following are the examples of *-inductive limits of t.v. spaces.

Example 1. Let (E,u) be a t.v. space, M a subspace of E and q_1 the canonical map of E onto E/M. Then E/M under its quotient topology is the *-inductive limit of $[(E,u): q_1]$.

Example 2. Let E be a vector space and $\{u_\alpha: \alpha \in A\}$ a family of vector topologies on E. Suppose that, for each $\alpha \in A$, i_α is the identity map of $(E,u\ )$ into E. If u is the lower bound of the u_α, (E,u) is the *-inductive limit of $\{[(E,u_\alpha): i_\alpha];\ \alpha \in A\}$.

Proposition 6. Let (E,u) be the *-inductive limit of $\{[(E_\alpha,u_\alpha): f_\alpha];\ \alpha \in A\}$. For each $\alpha \in A$, let V_α be a circled neighbourhood of 0 in E_α, and let

$$U = \bigcup_{\Phi} \sum_{\alpha\in\Phi} f_\alpha(V_\alpha),$$

the union being taken over all finite subsets Φ of A. Then U is a neighbourhood of 0 in (E,u).

If A is countable, then as V_α runs through a base of circled neighbourhoods of 0 in E_α, the sets {U} form a base of neighbourhoods of 0 in (E,u).

§2. Locally Convex Spaces.

Definition 7. (a) A non-empty subset A of a vector space E is called convex (semi-convex) if for all $x,y \in A$ and $0 \leq \lambda \leq 1$, $\lambda x + (1-\lambda)y \in A$ (respectively $A + A \subset \lambda A$ for some $\lambda > 0$).

(b) A t.v. space (E,u) is called a locally convex space (in short, an l.c. space) if there exists a neighbourhood basis of 0 in (E,u), consisting of convex sets.

(c) A t.v. space (E,u) is almost convex if each bounded subset of E is contained in some closed, circled, semiconvex and bounded set.

Definition 8. (a) Let E be a vector space. A nonnegative real-valued function ρ defined on E is said to be a gauge if

(i) $\rho(\lambda x) = \lambda\rho(x)$, for all $\lambda \geq 0$ and $x \in E$;

(ii) $\rho(x+y) \leq \rho(x) + \rho(y)$ for all $x,y \in E$.

Clearly $\rho(0) = 0$.

(b) A seminorm ρ on E is a gauge on E for which (i) is satisfied in the stronger form i.e.

(i)' $\rho(\lambda x) = |\lambda|\rho(x)$, $x \in E$ and $\lambda \in \mathbb{R}$.

(c) In addition, if $\rho(x) = 0$ implies $x = 0$, then ρ is called a norm and is denoted by $\|\cdot\|$.

A norm is always a seminorm but not conversely.

Let (E,u) be a t.v. space. There is a one-to-one correspondence between the sets of all closed, circled and convex subsets of E containing 0 as their interior point and the set of all continuous semi-norms defined on E. Since there exists a basis of closed, circled and convex neighbourhoods of 0 in an l.c. space (E,u), the topology u can also be defined by a subset of all continuous semi-norms on (E,u). An l.c. space (E,u) is metrizable iff u can be described by a countable family of continuous semi-norms. The set of all semi-norms on a vector space E defines a topology which is the finest l.c. topology. The subspaces, quotients and products of l.c. spaces are again l.c. spaces.

<u>Definition 9</u>. A t.v. space (E,u) is called a normed space if u can be defined by a norm.

A normed space is clearly an l.c. space. A t.v. space (E,u) is normable iff (E,u) possesses a bounded, convex neighbourhood of 0.

<u>Definition 10</u>. (a) Let $\{(E_\alpha, u_\alpha);\ \alpha \in A\}$ be a family of l.c. spaces, f_α a linear map of E_α into a vector space E for each α and $E = \bigcup_\alpha f_\alpha(E_\alpha)$. Let u be the finest l.c. topology on E for which each f_α is continuous. Then (E,u) is called the inductive limit of E_α's.

(b) The direct sum $E = \bigoplus_\alpha E_\alpha$ of a family $\{(E_\alpha, u_\alpha);\ \alpha \in A\}$ of l.c. spaces, endowed with the finest l.c. topology u such that the embedding $E_\alpha \to E$ is continuous for each α, is the inductive limit of E_α's. The topology u is called the direct sum topology.

A neighbourhood basis {U} at 0 in (E,u) is described as follows: A circled, convex and absorbing subset U of E is a neighbourhood of 0 in (E,u) iff for each α, $f_\alpha^{-1}(U)$ is a neighbourhood of 0 in (E_α, u_α).

Proposition 7. Let (E,u) be an inductive limit of a family $\{[(E_\alpha, u_\alpha): f_\alpha]: \alpha \in A\}$ of l.c. spaces and (F,v) any l.c. space. If f is a linear map of E into F, then f is continuous iff the composition map $f \circ f_\alpha$ is continuous for each α.

§3. The Hahn-Banach Theorem.

Definition 11. (a) The translate of a (vector) subspace M of a vector space E is called a linear manifold, i.e., a set F of the form $x_0 + M$, where $x_0 \in E$.

We also say that $x_0 + M$ is a linear manifold through x_0 parallel to M. Clearly $x_1 + M = x_0 + M$ for each $x_1 \in x_0 + M$.

(b) A maximal proper linear manifold is called a hyperplane.

The following is the geometric form of the Hahn-Banach theorem:

Theorem 4. Let (E,u) be a t.v. space, A a non-empty open convex subset of E and M a linear manifold which does not meet A. Then there exists a closed hyperplane H in E which contains M and does not meet A.

The Hahn-Banach theorem in its often-used analytic form is as follows:

Theorem 5. Let E be a vector space and ρ a semi-norm on E. Let M be a vector subspace of E and f a linear functional defined on M such that $|f(x)| \leq \rho(x)$ for $x \in M$. Then there exists a linear functional $\tilde{f}$ on E which coincides with f on M and $|\tilde{f}(x)| \leq \rho(x)$ for all $x \in E$.

The following are some conclusions drawn from the above theorem.

Proposition 8. Let E be an l.c. space, M a subspace of E and f a continuous linear functional on M. Then there exists a continuous linear functional $\tilde{f}$ defined on E such that $\tilde{f}(x) = f(x)$ for all $x \in M$.

Proposition 9. Let E be an l.c. space, M a closed subspace of E and $z \in E\backslash M$. Then there exists a continuous linear functional f on E such that $f(z) = 1$ and $f(x) = 0$ for $x \in M$.

Proposition 10. In an l.c. space (E,u), every closed linear manifold M is the intersection of the closed hyperplanes which contain it.

Proposition 11. Let E be an l.c. space, A a closed convex non-empty set in E and $x_0 \in E\backslash A$. Then there exists a continuous linear functional f defined on E and a real number α such that $f(x) > \alpha$ for all $x \in A$ and $f(x_0) \leq \alpha$.

Remark. If f is a linear functional on a vector space E, the subsets $\{x \in E: f(x) \geq \alpha\}$ and $\{x \in E: f(x) \leq \alpha\}$ are called closed half-spaces while $\{x \in E: f(x) > \alpha\}$ and $\{x \in E: f(x) < \alpha\}$ are called open half-spaces.

Corollary 3. In an l.c. space, every closed convex set A is the intersection of the closed half-spaces which contain it.

4. Completeness, quasicompleteness and sequential completeness.

Let E be any set. A non-empty family $\mathfrak{F}$ of subsets of E is called a filter if it satisfies:

(F_0) $\emptyset \notin \mathfrak{F}$;

(F_1) $A, B, \in \mathfrak{F}$ implies that $A \cap B \in \mathfrak{F}$;

(F_2) $A \in \mathfrak{F}$ and $A \subset B$ imply that $B \in \mathfrak{F}$.

A filter $\mathfrak{F}$ in a topological space (E,u) is said to be convergent to $x_0 \in E$ if each neighborhood of x_0 belongs to $\mathfrak{F}$.

In a t.v. space (E,u), a filter $\mathfrak{F}$ is called a Cauchy filter if for every neighbourhood V of 0 in (E,u) there exists a member $A \in \mathfrak{F}$ such that $A - A \subset V$.

Definition 12. (a) A t.v. space (E,u) is said to be complete if every Cauchy filter is convergent; more generally, a subset A of E is complete if every Cauchy filter in A is convergent to a point of A.

(b) A t.v. space (E,u) is said to be quasi-complete if every closed bounded subset of (E,u) is complete.

(c) A t.v. space (E,u) is said to be sequentially complete (or semi-complete) if every Cauchy sequence in E converges.

Every complete space is quasi-complete but not conversely, and every quasi-complete space is sequentially complete but not conversely. A complete metrizable l.c. space is called a Fréchet space and a complete normed space is called a Banach space. A Banach space is clearly a Fréchet space, but not conversely.

Theorem 6.

(i) Any scalar multiple of any closed subset of a complete set is complete.

(ii) Any finite union of complete sets is complete.

(iii) A complete subset of a Hausdorff space is closed.

(iv) A complete and precompact subset of a t.v. space is compact.

(v) The product of any family of quasicomplete spaces is quasicomplete.

§5. Continuous and Open Linear Maps.

Let (E,u) and (F,v) be two t.v. spaces and f a linear map of E into F. Then

(i) f is continuous (almost continuous) if for each neighbourhood V of 0 in (F,v) there exists a neighbourhood U of 0 in (E,u) such that $f^{-1}(V) \supset U$ (respectively, $\overline{f^{-1}(V)} \supset U$).

(ii) f is open (almost open) if for each neighbourhood U of

0 in (E,u), there exists a neighbourhood V of 0 in (F,v) such that $f(U) \supset V$ (respectively, $\overline{f(U)} \supset V$).

A continuous (open) linear map is almost continuous (almost open) but not conversely.

(iii) $G_f = \{(x,f(x)): x \in E\} \subset E \times F$ is called the graph of f. G_f is said to be closed if it is a closed subset of $E \times F$.

Theorem 7. Let (E,u) be a t.v. space and (F,v) an F-space (i.e. complete metrizable t.v. space).

(a) If f is an almost continuous linear map of E into F such that the graph G_f is closed, then f is continuous.

(b) If g is a continuous almost open linear map of F into E, then g is open. (cf: [12] for proofs.).

Theorem 8. Let E be a t.v. space and F an F-space. If f is an almost open linear map of F into E such that the graph of f is closed in $F \times E$, then f is open.

Proposition 12. Let E and F be any two t.v. spaces. Let f be a linear map of F into E such that f(F) is of the second category in E. Then f is almost open.

Corollary 4. Let E be a t.v. space and F an F-space. Let f be a linear continuous map of F into E such that f(F) is of the second category in E. Then f is open and onto.

It is known (lemma 2, § 1, Ch.II) that each linear map of a barrelled space into l.c. space is almost continuous and each linear map of an l.c. space onto a barrelled space is almost open. (cf. 12).

Definition 12'. An l.c. space is said to be B-complete if each continuous almost open linear map onto any l.c. space is open.

Theorem 8'. Let F be a barrelled space and E B-complete. Then

(i) (O.M.) Each continuous linear map of E onto F is open.

(ii) (C.G.) Each linear map of F into E with closed graph is continuous.

Proof: See [12].

6. $\mathfrak{S}$-topologies.

Let (E,u) and (F,v) be two t.v. spaces. We write $\mathcal{L}(E,F)$ to denote the vector space of all continuous linear maps of E into F.

Let $\mathfrak{S}$ be a family of subsets of E. We define a topology in $\mathcal{L}(E,F)$ of uniform convergence over the sets M in $\mathfrak{S}$ as follows: Let V be a neighbourhood of 0 in F and $M \in \mathfrak{S}$. Let

$$T(M,V) = \{f \in \mathcal{L}(E,F): \ f(M) \subset V\}.$$

The family of all T(M,V), as M runs over $\mathfrak{S}$ and V over the neighbourhoods of 0 in F, forms a subbasis for a topology on $\mathcal{L}(E,F)$ called the $\mathfrak{S}$-topology. The vector space $\mathcal{L}(E,F)$ with an $\mathfrak{S}$-topology is denoted by $(\mathcal{L}(E,F),\mathfrak{S})$.

(1) $\mathcal{L}(E,F)$, equipped with an $\mathfrak{S}$-topology, is a t.v. space iff f(M) is bounded in F for each $M \in \mathfrak{S}$ and $f \in \mathcal{L}(E,F)$. If, in addition, F is an l.c. space, the $\mathfrak{S}$-topology is also locally

convex and so $\mathcal{L}(E,F)$ is an l.c. space. If $\mathfrak{S}$ consists of bounded sets of E such that $\cup M$ is total in E and F is an l.c. space, then the $\mathfrak{S}$ -topology is Hausdorff.

The most important particular cases of the $\mathfrak{S}$-topology are as follows:

(a) $\mathfrak{S}$ is the family of all finite subsets of E. In this case, the $\mathfrak{S}$-topology is called the topology of simple convergence.

(b) $\mathfrak{S}$ is the family of all compact (precompact) subsets of E. The $\mathfrak{S}$-topology is then called the topology of compact (precompact) convergence.

(c) $\mathfrak{S}$ is the family of all bounded subsets of E. The $\mathfrak{S}$-topology, in this case, is called the topology of uniform convergence on bounded sets, or strong topology.

If E and F are normed spaces, then the topology of uniform convergence on bounded sets is the normed topology given by : $\|f\| = \sup_{\|x\| \leq 1} \|f(x)\|$.

(2) Let (E,u) and (F,v) be two l.c. spaces such that (F,v) is complete. Then every continuous linear map f of E into F can be extended uniquely to a continuous linear map $\tilde{f}$ of $\tilde{E}$ into F, where $\tilde{E}$ is the completion of E. Thus $\mathcal{L}(E,F)$ and $\mathcal{L}(\tilde{E},F)$ are algebraically isomorphic. The $\mathfrak{S}$-topology on $\mathcal{L}(E,F)$ may however be different from the $\mathfrak{S}$-topology on $\mathcal{L}(\tilde{E},F)$. But the $\mathfrak{S}$-topology on $\mathcal{L}(E,F)$ remains unchanged if one replaces $\mathfrak{S}$ by $\mathfrak{S}'$, where $\mathfrak{S}'$ consists of $\overline{M}$ (closure in E), $M \in \mathfrak{S}$.

Definition 13. A subset H of $\mathcal{L}(E,F)$ is said to be $\mathfrak{S}$-bounded if H is bounded in the $\mathfrak{S}$-topology.

In particular, H is simply bounded if H is bounded in the topology of simple convergence on $\mathcal{L}(E,F)$.

Proposition 13. Let H be a subset of $\mathcal{L}(E,F)$. Then the following statements are equivalent:

(a) H is bounded for the $\mathfrak{S}$-topology.

(b) For each neighbourhood V of 0 in F, $\bigcap_{f \in H} f^{-1}(V)$ absorbs each $M \in \mathfrak{S}$.

(c) For each $M \in \mathfrak{S}$, $\bigcup_{f \in H} f(M)$ is bounded in F.

Definition 14. Let H be a subset of $\mathcal{L}(E,F)$. H is said to be equicontinuous at 0 if for each neighbourhood V of 0 in F, there exists a neighbourhood U of 0 in E such that $f(U) \subset V$ for all $f \in H$.

Proposition 14. H is equicontinuous iff it is equicontinuous at 0.

Proposition 15. Let H be a subset of $\mathcal{L}(E,F)$. Then the following statements are equivalent:

(a) H is equicontinuous.

(b) For each neighbourhood V of 0 in F, $\bigcap_{f \in H} f^{-1}(V)$ is a neighbourhood of 0 in E.

(c) For each neighbourhood V of 0 in F, there exists a neighbourhood U of 0 in E such that $\bigcup_{f \in H} f(U) \subset V$.

Corollary 5. Each equicontinuous subset of $\mathcal{L}(E,F)$ is bounded for every $\mathfrak{S}$-topology, provided each $M \in \mathfrak{S}$ is bounded.

Proposition 16. Let (E,u) and (F,v) be two l.c. spaces and H an equicontinuous subset of $\mathcal{L}(E,F)$. The restrictions to H of the following topologies are identical.

(a) The simple convergence topology,

(b) The topology of precompact convergence.

Corollary 6. Let (E,u), (F,v) and H be as in Proposition 16. If $\mathfrak{F}$ is a filter on H which converges to a map f_0: $E \to F$ in the simple convergence topology, then $f_0 \in \mathcal{L}(E,F)$ and converges to f_0 uniformly on each precompact subset of E.

Lemma 1. Let E be a metrizable t.v. space and F any t.v. space. Let H be a family of linear maps from E into F. H is equicontinuous iff for each $x_0 \in E$ and each sequence $\{x_n\}$ in E converging to x_0, $\{f(x_n)\}$ converges to $f(x_0)$ uniformly in $f \in H$.

§7. Duality Theory.

Let (E,u) be an l.c. space. The vector space $E' = \mathcal{L}(E,\mathbb{R})$ is called the topological dual (or simply, dual) of E.

The value of a linear functional $x' \in E'$ at $x \in E$ is denoted by $\langle x,x' \rangle$. Clearly the mapping $(x,x') \to \langle x,x' \rangle$ of $E \times E'$ into $\mathbb{R}$ is bilinear (i.e. linear in each variable). $\langle E,E' \rangle$ is called a dual pair.

Definition 15. (a) The coarsest l.c. topology for which the map $x \to \langle x,x' \rangle$ for each $x' \in E'$ is continuous, is called the weak topology $\sigma(E,E')$ on E.

(b) The coarsest l.c. topology for which the map $x' \to \langle x,x' \rangle$ for each $x \in E$ is continuous is called the weak*-topology $\sigma(E',E)$ on E'.

Remark 3. $\sigma(E',E)$ is precisely the topology of simple convergence on E' induced from $\mathbb{R}^E$.

Definition 16. (a) For each subset A of an l.c. space (E,u), $A^o = \{x' \in E': \langle x,x' \rangle \leq 1 \text{ for all } x \in A\}$ is called the polar of A.

(b) $A^{oo} = \{x \in E: \langle x,x' \rangle \leq 1 \text{ for all } x' \in A^o\}$ is called the bipolar of A.

Proposition 17. Let A,B and $\{A_\alpha : \alpha \in A\}$ be subsets of an l.c. space E.

(i) If $A \subset B$ then $A^o \supset B^o$.

(ii) If $\lambda \neq 0$, then $(\lambda A)^o = \lambda^{-1} A^o$. In particular, A^o is absorbing iff A is $\sigma(E,E')$-bounded.

(iii) A^o is a $\sigma(E',E)$-closed, circled, convex set.

(iv) A^{oo} is equal to the convex, $\sigma(E,E')$-closure of $A \cup \{0\}$.

(v) $A^{ooo} = A^o$.

(vi) $(\bigcup_\alpha A_\alpha)^o = \bigcap_\alpha A_\alpha^o$.

(vii) If each A_α is a $\sigma(E,E')$-closed convex subset of E containing 0, then $(\bigcap_\alpha A_\alpha)^0$ is the $\sigma(E',E)$-closed convex hull of $\bigcup_\alpha A_\alpha^0$.

(viii) If A is a subspace of E, A^0 is a $\sigma(E',E)$-closed subspace of E'.

(ix) Let A be a subspace of E. $A^{00} = A$ iff A is $\sigma(E,E')$-closed.

Let A be a subspace of E. Then A^0 is a $\sigma(E',E)$-closed subspace of E'. Consider E'/A^0 with the quotient topology. For each $x^0 \in E'/A^0$, $x^0 = x' + A^0$ $(x' \in E')$ and therefore, for each $x \in A$, $\langle x,x_1^0\rangle = \langle x,x_2^0\rangle$ if $x_1' - x_2' \in A^0$. Clearly, $(x,x^0) \to \langle x,x^0\rangle$ is a bilinear map of $A \times (E'/A^0)$. Therefore A and E'/A^0 are duals of each other.

<u>Proposition 18</u>. Let A be a subspace of E and E' the dual of E. Then:

(i) The weak topology $\sigma(A,E'/A^0)$ on A is the topology induced by $\sigma(E,E')$ on A.

(ii) For the topology $\sigma(E'/A^0,A)$ to be equal to the quotient w*-topology on E'/A^0, it is necessary and sufficient that A be $\sigma(E,E')$-closed in E.

<u>Proposition 19</u>. Let E_α $(\alpha \in A)$ be an l.c. space and E_α' its dual. Then

(a) $(\prod_\alpha E_\alpha)' = \sum_\alpha E_\alpha'$,

(b) $(\sum_\alpha E_\alpha)' = \prod_\alpha E_\alpha'$.

Proposition 20. Let (E,u) be an l.c. space and $M' \subset E'$. M' is equicontinuous iff there exists a neighbourhood V of 0 in (E,u) such that $M' \subset V^{\circ}$ (or iff M'° is a neighbourhood of 0 in E).

Proposition 21. Let (E,u) be an l.c. space and let $\mathfrak{S}$ be the family of all equicontinuous subsets of E'. Then u is equal to the $\mathfrak{S}$-topology.

Theorem 9. (Alaoglu-Bourbaki). Let (E,u) be an l.c. space. Then any equicontinuous subset of E' is $\sigma(E',E)$-relatively compact.

Let u and v be two l.c. topologies on a vector space E. u and v are compatible with the duality (in notation $u \sim v$) if (E,u) and (E,v) have the same topological dual.

For any l.c. space (E,u), $\sigma(E,E') \sim u$.

Proposition 22. Let u and v be two l.c. topologies on a vector space E which are compatible with duality. Then a convex subset A of E is u-closed iff it is v-closed.

Definition 17. Let (E,u) be an l.c. space and let $\mathfrak{S}$ be the family of all circled, convex, and $\sigma(E',E)$-compact subsets of E'.

(a) The $\mathfrak{S}$-topology on E is called the Mackey topology and is denoted by $\tau(E,E')$.

(b) (E,u) is called a Mackey space if $u = \tau(E,E')$.

Theorem 10. An l.c. space (E,u) is a Mackey space iff each convex, $\sigma(E',E)$-relatively compact subset of E' is equicontinuous.

Theorem 11. (Mackey-Arens). The Mackey topology $\tau(E',E)$ (respectively $\tau(E,E')$) is the finest l.c. topology on E' (respectively, on E) compatible with the dual pair $\langle E,E'\rangle$.

Let (E,u) be an l.c. space and let $\mathfrak{S}$ be the family of all bounded subsets of E. Then the $\mathfrak{S}$-topology on E' is called the strong topology and is denoted by $\beta(E',E)$.

Proposition 23. Let (E,u) be an l.c. space. Then (a) every equicontinuous subset of E' is $\beta(E',E)$-bounded; (b) every convex subset of E' which is $\sigma(E',E)$-relatively compact is $\beta(E',E)$-bounded.

Theorem 12. (Banach-Mackey). Every circled, convex, closed, bounded and sequentially complete subset of an l.c. space (E,u) is $\beta(E,E')$-bounded.

Corollary 7. The bounded subsets of an l.c. space E are the same for all the l.c. topologies compatible with the dual pair $\langle E,E'\rangle$.

Proposition 24. Let (E,u) be a sequentially complete l.c. space. Then every $\sigma(E,E')$-bounded subset of E ($\sigma(E',E)$-bounded subset of E') is $\beta(E,E')$-bounded (respectively, $\beta(E',E)$-bounded).

Theorem 13. (Eberlein). Let (E,u) be a quasicomplete l.c. space. If every sequence in a subset A of E has a weak cluster point, then the weak closure of A is weakly compact.

Corollary 8. Let (E,u) be an l.c. space and let A be a subset of E such that every sequence of points of A has a cluster point. If $(E, \tau(E,E'))$ is quasicomplete then A is relatively compact.

We call $E'^{\beta} = (E', \beta(E',E))$, the strong dual of E'. The set E' without the mention of any topology will be understood to have been equipped with $\sigma(E',E)$. The dual $E'^{\beta\prime}$ of E'^{β} is called the bidual of E.

Definition 18. (a) If $E'^{\beta\prime} = E$ (algebraically), then E is said to be semi-reflexive.

(b) If $E'^{\beta,\beta} = (E,u)$ (topologically), then E is said to be reflexive.

Clearly every reflexive space is semi-reflexive.

Theorem 14. An l.c. space (E,u) is semi-reflexive iff it is $\sigma(E,E')$-quasi-complete.

Proposition 25. Let (E,u) be a reflexive space. Then every closed, circled, convex and absorbing set in E is a neighbourhood of 0.

Let (E,u) be a metrizable l.c. space. The topology u can always be given by a decreasing sequence $V_1 \supset V_2 \supset \ldots$ of circled convex neighbourhoods of 0.

Proposition 26. Let (E,u) be a metrizable l.c. space. If $V_1 \supset V_2 \supset, \ldots,$ is a neighbourhood basis of 0 in (E,u), then the sets $\{V_n^o;\ n \geq 1\}$ form a fundamental sequence of $\beta(E',E)$-bounded sets in E'.

Proposition 27. Let (E,u) be a metrizable l.c. space. If $\{H_n\}$ is a sequence of equicontinuous subsets of E" such that $H = \bigcup_{n=1}^{\infty} H_n$ is $\beta(E",E')$-bounded, then H is equicontinuous.

Equivalently, if $\{V_n\}$ is a sequence of circled, convex and closed neighbourhoods of 0 in $(E', \beta(E',E))$ such that $V = \bigcap_{n=1}^{\infty} V_n$ is bornivorous, then V is a neighbourhood of 0 in $(E',\beta(E',E))$.

§8 Topological Tensor Products.

Let E, F and G be vector spaces. A map h of E × F into G is called bilinear if for each $x \in E$ and each $y \in F$, the partial maps h_x and h_y defined by $h_x(y) = h(x,y)$ and $h_y(x) = h(x,y)$ are linear. Furthermore, if $G = \mathbb{R}$, then f is called a bilinear functional.

Let B(E,F) denote the vector space of all bilinear functionals on E × F, and B*(E,F) its algebraic dual. The map ϕ of E × F into B*(E,F), defined by

$$\phi(x,y)(h) = h(x,y)$$

for $(x,y) \in E \times F$ and $h \in B(E,F)$, is bilinear. The image $\phi(E \times F)$ of E × F is in general not a vector space; the linear hull of $\phi(E \times F)$ in B*(E,F) is called the tensor product of E and F and is denoted by $E \otimes F$; ϕ is called the canonical bilinear map of E × F into $E \otimes F$. We write $x \otimes y$ to represent $\phi(x,y)$; each element $h \in E \otimes F$ can be written as a finite sum of the form $\Sigma(x_i \otimes y_i)$, $x_i \in E$, $y_i \in F$; but this representation of h is not unique.

Proposition 28. Let E, F and G be vector spaces and let ϕ be the canonical map of $E \times F$ into $E \otimes F$. The map $h \to h \circ \phi$ is an isomorphism of the vector space $L(E \otimes F, G)$ of linear maps of $E \otimes F$ into G onto the vector space $B(E \times F; G)$ of bilinear maps of $E \times F$ into G.

Corollary 9. The algebraic dual of $E \otimes F$ can be identified with $B(E,F)$; under this identification, $L(E \otimes F, \mathbb{R})$ can be identified with $B(E \times F; \mathbb{R})$.

Proposition 29. Let E and F be two l.c. spaces and ϕ the canonical map of $E \times F$ into $E \otimes F$.

(a) There is a finest l.c. topology u_p, called the projective (tensor product) topology, on $E \otimes F$ under which ϕ is continuous.

(b) If η and ζ are neighbourhood bases at 0 in E and F respectively, the circled convex hulls of the sets $\phi(U,V)$, $U \in \eta$, $V \in \zeta$, form a neighbourhood basis for the topology u_p.

(c) A linear map f of $(E \otimes F, u_p)$ into an l.c. space G is continuous iff the bilinear map $f \circ \phi$ of $E \times F$ into G is continuous.

Proposition 30. The topological tensor product $(E \otimes F, u_p)$ is Hausdorff iff both E and F are.

Let E and F be two l.c. spaces. The finest l.c. topology u_i on $E \otimes F$ for which the canonical map ϕ is separately continuous (i.e., all partial maps ϕ_x and ϕ_y are continuous), is called the inductive (tensor product) topology; it is the inductive topology in the sense of §2, Definition 5.

§9. Topological bases.

Definition 19. Let (E,u) be a t.v. space and E* its algebraic dual. Let $\{x_i\}$ and $\{f_i\}$ be sequences in E and E* respectively. $\{x_i,f_i\}$ is called a biorthogonal system for E if $f_i(x_j) = \delta_{ij}$, where δ_{ij} is the Kronecker delta.

Definition 20. A sequence $\{x_i\}$ in a t.v. space E is called a basis for E if for each $x \in E$ there is a unique sequence $\{\alpha_i\} \subset \mathbb{R}$ such that $x = \lim_n \sum_{i=1}^{n} \alpha_i x_i$ in the topology of E.

Clearly, each coefficient α_i, $f_i(x) = \alpha_i$, defines a linear functional f_i on E. However, the coefficient functionals f_i need not be continuous. If each f_i is continuous, then $\{x_i\}$ is called a Schauder basis for E.

In a Banach (or even Fréchet) space E, every basis is a Schauder basis.

Definition 21. Let E be a t.v. space, E' its topological dual and let Λ be an index set of arbitrary cardinality. The double family $\{x_\lambda, f_\lambda\}$ where $x_\lambda \in E$ and $f_\lambda \in E'$ for all $\lambda \in \Lambda$ is called a biorthogonal system if $f_\lambda(x_\mu) = \delta_{\lambda\mu}$.

(a) $\{x_\lambda, f_\lambda\}$ is maximal with respect to E if there is no biorthogonal system which contains $\{x_\lambda, f_\lambda\}$ properly.

(b) $\{x_\lambda, f_\lambda\}$ is a generalized basis for E if $f_\lambda(x) = 0$, for all $\lambda \in \Lambda$ implies $x = 0$.

(c) If the set of basis elements $\{x_\lambda\}$ of a biorthogonal system $\{x_\lambda, f_\lambda\}$ in E is total in E, then $\{x_\lambda, f_\lambda\}$ is called a dual generalized basis for E. Moreover, if such a basis is also a generalized basis for E, it is called a Markuschevich (an extended Markuschevich) basis if Λ is countable (or not).

<u>Theorem 15</u>.

(i) Every Schauder basis for E is a Markuschevich basis;

(ii) A generalized basis for E is a maximal biorthogonal system with respect to E.

<u>Theorem 16</u>. Let (E,u) be an l.c. space. Then every weak (extended) Markuschevich basis for E is a (an extended) Markuschevich basis for E.

<u>Definition 22</u>. Let (E,u) and (F,v) be t.v. spaces. A sequence $\{x_i\}$ in E is said to be similar to a sequence $\{y_i\}$ in F if for all sequences $\{a_i\} \subset \mathbb{R}$, $\sum_{i=1}^{\infty} a_i x_i$ converges (in E) iff $\sum_{i=1}^{\infty} a_i y_i$ converges (in F).

Let $D(\Lambda)$ be the vector space of all real-valued functions on Λ in which a function Z is zero iff $Z(\lambda) = 0$ for every $\lambda \in \Lambda$. The map $\Phi: E \to D(\Lambda)$ defined by $\Phi(x) = \{f_\lambda(x)\}$ is called the coefficient map.

<u>Definition 23</u>. Let E and F be t.v. spaces. Let $\{x_\lambda\}_{\lambda\varepsilon\Lambda}$ and $\{y_\lambda\}_{\lambda\varepsilon\Lambda}$ be generalized bases in E and F respectively. Then $\{x_\lambda\}$ and $\{y_\lambda\}$ are similar if there exist families $\{f_\lambda\}$ and $\{g_\lambda\}$ of coefficient functionals for $\{x_\lambda\}$ and $\{y_\lambda\}$ respectively such that $\Psi(E) = \Phi(F)$, where Ψ and Φ are the coefficient maps determined by each family of coefficient functionals.

Let (F,v) be an l.c. space and let $z, z_\mu \in F$, $\mu \in \Lambda$. Denote by Φ the set of all finite subsets of Λ. The family $\{z_\mu; \mu \in \Lambda\}$ is said to be unconditionally Cauchy if for each v-neighbourhood U of 0, there exists $\phi_1 \in \Phi$ such that $\sum_{\mu\varepsilon\phi} z_\mu \in U$ for all $\phi \in \Phi$ with $\phi \cap \phi_1 = \emptyset$, and to be unconditionally convergent to z if for each such U, there exists $\phi_2 \in \Phi$ such that $z - \sum_{\mu\varepsilon\phi} z_\mu \in U$ for all $\phi \in \Phi$ with $\phi_2 \subset \phi$. In the latter case, we write $z = \sum_{\mu\varepsilon\Lambda} z_\mu$.

I. If $\Lambda' \subset \Lambda$ and $\{z_\mu: \mu \in \Lambda\}$ is unconditionally Cauchy, so is $\{z_\mu: \mu \in \Lambda'\}$.

II. If $\{z_\mu: \mu \in \Lambda\}$ is unconditionally Cauchy and $\{\alpha_\mu: \mu \in \Lambda\}$ is a bounded family of scalars, then $\{\alpha_\mu z_\mu: \mu \in \Lambda\}$ is unconditionally Cauchy.

III. If $\{z_\mu: \mu \in \Lambda\}$ is unconditionally convergent, it is also unconditionally Cauchy. Conversely, if F is quasicomplete and $\{z_\mu: \mu \in \Lambda\}$ is unconditionally Cauchy, it is also unconditionally convergent.

IV. If F is metrizable and $\{z_\mu : \mu \in \Lambda\}$ is unconditionally Cauchy, then $\{\mu: z_\mu \neq 0\}$ is at most countable.

V. $\{z_\mu : \mu \in \Lambda\}$ is unconditionally Cauchy iff $\{\sum_{\mu\in\phi} z_\mu : \phi \in \Phi\}$ is precompact (if $\phi = \emptyset$, $\sum_{\lambda\in\phi} z_\lambda$ is defined to be 0).

VI. If for each countable subfamily $\{z_{\mu(n)} : n \in \mathbb{N}\}$ of $\{z_\mu : \mu \in \Lambda\}$, Λ infinite, $\Sigma z_{\mu(n)}$ is $\sigma(F,F')$-convergent, then $\{z_\mu : \mu \in \Lambda\}$ is unconditionally Cauchy for all topologies of the dual pair $\langle F,F'\rangle$.

§10. Ordered topological vector spaces.

We call a vector space E an "ordered vector space" if it is equipped with a reflexive, transitive and antisymmetric relation $\leq$ which satisfies the following conditions:

(a) $x \leq y$ implies $x + z \leq y + z$ for all $x,y,z \in E$.

(b) $x \leq y$ implies $\alpha x \leq \alpha y$ for all $x,y \in E$ and $\alpha > 0$.

The set $C = \{x \in E: x \geq 0\}$ is called the positive cone (or simply, the cone). Clearly

(a_1) $C + C \subset C$

(b_1) $\lambda C \subset C$, for $\lambda > 0$

(c_1) $C \cap (-C) = \{0\}$.

On the other hand, if C is a subset of a vector space E satisfying (a_1), (b_1) and (c_1), then $x \leq y$ iff $y - x \in C$ defines an order relation $\leq$ on E and thus E is an ordered vector space with C as its positive cone.

A set C satisfying only (a_1) and (b_1) and containing 0 is called a wedge. We write (E,C) to denote an ordered vector space with C as its positive cone.

Let (E,C) be an ordered vector space. If $x \leq y$, $x,y \in E$, then the set $[x,y] = \{z \in E: \ x \leq z \leq y\}$ is called the order interval between x and y. A subset A of E is said to be order-bounded if $A \subseteq [x,y]$, for some $x,y \in E$. A subset B of E is called order-bornivorous if it absorbs all order-bounded subsets of E. The cone C in E is called generating if $E = C-C$. An element $e \in E$ is called an order unit if for each $x \in E$, there is an $\alpha > 0$ such that $x \leq \alpha e$. Clearly C is generating if E contains an order unit.

Let (E,C) be an ordered vector space and M a vector subspace of E. Then $(M, C \cap M)$ is an ordered vector space. If g is the canonical map of E onto E/M, then g(C) is a wedge in E/M; it need not be a cone.

If $\{(E_\alpha, C_\alpha): \alpha \in A\}$ is a family of ordered vector spaces, then $(\prod_\alpha E_\alpha, \prod_\alpha C_\alpha)$ and $(\bigoplus_\alpha E_\alpha, \bigoplus_\alpha C_\alpha)$ are ordered vector spaces.

Let A be a subset of an ordered vector space (E,C) and let $x \in E$ be such that

(a_2) $a \leq x$ for all $a \in A$, and

(b_2) $a \leq z$ for all $a \in A$ implies $x \leq z$.

Then x is called the supremum of A, written $x = \sup(A)$. Dually, we have infimum of A, written inf(A). For every pair $x,y \in E$, if the $\sup\{x,y\} = x \vee y$ and $\inf(x,y) = x \wedge y$ exist, then (E,C)

is called a vector lattice. Since $x \vee y = -\{(-x) \wedge (-y)\}$, (E,C) is a vector lattice if the supremum (or infimum) of every pair of elements in E exists.

Let (E,C) be a vector lattice and $x \in E$. We define $x^+ = \sup\{x,0\}$, $x^- = (-x)^+$, $|x| = \sup\{x,-x\}$, which are respectively called the positive part, the negative part and the absolute value of x. It can easily be shown that

$$x = x^+ - x^- \text{ and } |x| = x^+ + x^-.$$

We conclude from $x = x^+ - x^-$ that the cone in a vector lattice is always generating. A vector lattice (E,C) always satisfies the following so-called decomposition property:

(*) $[0,x] + [0,y] = [0,x+y]$ for all $x,y \in C$.

Let (E,C) be an ordered vector space and B a subset of E. B is called order complete if every directed ($\leq$) subset D of B that is majorized in E has a supremum belonging to B. An order complete vector space (E,C) is a vector lattice iff C is generating.

Let (E,C) be a vector lattice and B a subset of E. We call B a solid set if $|x| \leq |y|$ for $y \in B$ implies that $x \in B$. A solid subspace M of E is called a lattice ideal of E. If (E,C) is an order complete vector lattice, a lattice ideal M in E is called a band in E if M contains the supremum of every subset of M that is majorized in E.

Let the vector space $\omega = \{x = (x_n): x_n \in \mathbb{R}\}$ of all sequences be ordered by the cone

$$C = \{x = (x_n) \in \omega: \; x_n \geq 0 \text{ for all } n \geq 1\}.$$

Then (ω, C) is a vector lattice. We denote by ϕ the vector subspace of ω consisting of those elements of ω which have finitely many non-zero components.

A vector subspace λ of ω such that $\phi \subset \lambda$ is called a sequence space. The Köthe dual $\lambda^\times$ of a sequence space λ is defined as follows:

$$\lambda^\times = \{t = (t_n) \in \omega: \; \sum_{n=1}^{\infty} |t_n x_n| < \infty \text{ for all } x = (x_n) \in \lambda\}.$$

$\lambda^\times$ is also a sequence space and $\lambda \subset \lambda^{\times\times}$.

If $\lambda = \lambda^{\times\times}$, then λ is called a perfect sequence space.

Let E be a locally compact, σ-compact Hausdorff space and μ a non-negative Radon measure on E. A real-valued μ-measurable function f on E is called locally integrable if

$$\int_K |f| \, d\mu < \infty$$

for each compact subset K of E.

We define an equivalence relation on the class of all locally integrable functions on E as follows:

Two locally integrable functions f and g are equivalent if $\int_K |f-g| \, d\mu = 0$ for each compact subset K of E.

The family Ω of all equivalence classes with respect to this equivalence relation is a vector space. Let $A \subset \Omega$. We define $\Lambda = \{f \in \Omega: \int_E |fg| d\mu < \infty \;$ for all $g \in A\}$, and

$\Lambda^{\times} = \{h \in \Omega: \int_E |fh|\, d\mu < \infty$ for all $f \in \Lambda\}$.

Λ is called a Köthe function space and $\Lambda^{\times}$ its Köthe dual. Clearly $\Lambda = \Lambda^{\times\times}$. The vector spaces Λ and $\Lambda^{\times}$ are put in duality by the bilinear functional:

$$\langle f,g \rangle = \int_E fg\, d\mu, \quad \text{for } f \in \Lambda,\ g \in \Lambda^{\times}.$$

Definition 24. An ordered vector space which is also a t.v. space is called an ordered t.v. space.

We write (E,C,u) to denote an ordered t.v. space with the topology u.

Let A be a subset of an ordered vector space (E,C). The full hull [A] of A is defined by:

$$[A] = (A+C) \cap (A-C) = \{z \in E: x \leq z \leq y, \text{ for } x,y \in A\}.$$

Clearly $A \subset [A]$. If A = [A], then A is said to be full. The full hull of a convex (circled) set is convex (respectively circled).

Definition 25. Let (E,C,u) be an ordered t.v. space. The positive cone C is said to be normal for u if there exists a neighbourhood basis η of 0 in (E,C,u) consisting of full sets.

Remark 4. (a) The members of η can be chosen to be circled.
(b) If u is an l.c. topology, we can assume that members of η are convex.

Theorem 17. Let (E,C,u) be an ordered t.v. space. The following statements are equivalent:

(i) C is normal for u.

(ii) There is a neighbourhood basis $\eta = \{V\}$ of 0 for u such that $0 \le x \le y \in V$ implies $x \in V$.

(iii) For any two nets $\{x_\alpha: \alpha \in A\}$ and $\{y_\alpha: \alpha \in A\}$ in (E,C,u) if $0 \le x_\alpha \le y_\alpha$ for all $\alpha \in A$, and if $\{y_\alpha: \alpha \in A\}$ converges to 0 for u, then $\{x_\alpha: \alpha \in A\}$ converges to 0 for u.

(iv) Given an u-neighbourhood V of 0, there exists an u-neighbourhood U of 0 such that $0 \le x \le y \in U$ implies $x \in V$.

Proposition 31. Let (E,C,u) be an ordered t.v. space. If C is normal for u, then every order-bounded subset of E is u-bounded.

Proposition 32. Let (E,u) be a t.v. space ordered by a normal cone C with non-empty interior. Then (E,u) is normable.

Theorem 18. Let (E,C,u) be an ordered t.v. space such that (E,C) is a vector lattice. The following statements are equivalent:

(i) $(x,y) \to x \wedge y$ is a continuous map of $E \times E$ into E.

(ii) $(x,y) \to x \vee y$ is a continuous map of $E \times E$ into E.

(iii) $x \to x^+$ is a continuous map on E.

(iv) $x \to x^-$ is a continuous map on E.

(v) $x \to |x|$ is a continuous map on E.

If the cone C is normal, the continuity of each of the maps in (iii), (iv) and (v) is equivalent to its continuity at 0. Sometimes the maps in (i) - (v) are referred to as lattice operations.

Definition 26. Let (E,C,u) be an ordered t.v. space such that (E,C) is a vector lattice. (E,C,u) is called a t.v. lattice if there is a neighbourhood basis at 0 for u consisting of solid sets. u is called a locally solid linear topology.

Theorem 19. Let (E,C,u) be an ordered t.v. space such that (E,C) is a vector lattice. Then (E,C,u) is a t.v. lattice iff C is normal and the lattice operations are continuous.

Proposition 33. Let (E,C,u) be a t.v. lattice.

(a) Let B be a subset of E. If B is solid, so is its closure $\overline{B}$.

(b) There exists a neighbourhood basis of 0 consisting of closed and solid sets.

(c) If M is a lattice ideal in (E,u) so is its closure $\overline{M}$.

Proposition 34. Let (E,C,u) be a t.v. lattice.

(a) If M is a vector sublattice of E, then M is a t.v. lattice under the induced topology.

(b) If M is a lattice ideal in E, then E/M is a t.v. lattice for the quotient topology.

Let E be a non-empty set and $\mathcal{C}$ a family of subsets of E. A sub-family $\mathcal{C}_0$ of $\mathcal{C}$ is called a fundamental system for $\mathcal{C}$ if each member of $\mathcal{C}$ is contained in some member of $\mathcal{C}_0$.

Let (E,C,u) be an ordered t.v. space and $\mathcal{B}$ the family of all u-bounded subsets of E. C is called a $\mathcal{B}$-cone if

$\{\overline{A \cap C - A \cap C};\ A \in \mathfrak{B}\}$ is a fundamental system for $\mathfrak{B}$. C is a strict $\mathfrak{B}$-cone if $\{A \cap C - A \cap C;\ A \in \mathfrak{B}\}$ is a fundamental system for $\mathfrak{B}$.

Proposition 35. Let (E,C,u) be an ordered t.v. space such that (E,C) is a vector lattice. If the lattice operations are continuous, then C is a closed strict $\mathfrak{B}$-cone.

§11. Ordered locally convex spaces.

Definition 27. An ordered t.v. space (E,C,u) is called an ordered l.c. space if u is an l.c. topology.

Theorem 20. Let (E,C,u) be an ordered l.c. space. The following statements are equivalent:

(a) C is normal for u.

(b) There is a family $\{p_\alpha : \alpha \in A\}$ of seminorms generating the topology u such that $0 \leq x \leq y$ implies $p_\alpha(x) \leq p_\alpha(y)$ for all $\alpha \in A$. (Equivalently $p_\alpha(t+s) \geq p_\alpha(t)$ for all $t,s \in C$ and $\alpha \in A$).

Corollary 10. Let (E,C,u) be an ordered l.c. space with normal cone C. Then the closure $\overline{C}$ of C is also normal.

Proposition 36. (a) Let $\{(E_\alpha, C_\alpha, u_\alpha) : \alpha \in A\}$ be a family of ordered l.c. spaces and let $E = \bigoplus_\alpha E_\alpha$. Then $C = \bigoplus_\alpha C_\alpha$ is a normal cone in E iff C_α is normal in E_α.

(b) E is a l.c. vector lattice if each $(E_\alpha, C_\alpha, u_\alpha)$ is so.

Proposition 37. Let (E,C,u) be an ordered l.c. space with normal cone C. Then each continuous linear functional on E is the difference of two positive continuous linear functionals on E.

Theorem 21. Let (E,C,u) and (F,K,v) be ordered t.v. spaces with K normal in F. Then each positive linear map of E into F is continuous if one of the following conditions is satisfied:

(a) C has a non-empty interior.

(b) (E,u) is a bornological space, C is a sequentially complete strict $\mathcal{B}$-cone and (F,v) is an l.c. space.

(c) (E,u) is a metrizable t.v. space of second category which is ordered by a complete generating cone C and (F,v) is an l.c. space.

Corollary 11. Let (E,C,u) be an ordered t.v. space. Each positive linear functional on E is continuous if one of the following conditions is satisfied:

(a) C has a non-empty interior.

(b) (E,u) is a bornological space and C a sequentially complete strict $\mathcal{B}$-cone.

(c) (E,u) is a metrizable t.v. space of second category ordered by a complete generating cone.

Let (E,C,u) be an ordered l.c. space. An u-basis $\{x_n, f_n\}$ is called a positive u-basis if $\{x_n\} \subset C$ and all the f_n's are positive. Clearly a positive u-basis $\{x_n, f_n\}$ is an u-Schauder basis if any one of the conditions (a) - (c) of Corollary 11 is satisfied.

Definition 28.

(a) A t.v. lattice (E,C,u) is called an l.c. vector lattice if u is an l.c. topology.

In an l.c. vector lattice (E,C,u) there exists a neighbourhood basis of 0 consisting of closed, convex and solid sets. u is called a locally solid topology.

(b) A vector lattice (E,C) equipped with a norm $\|\cdot\|$ is called a normed vector lattice if $|x| \leq |y|$ implies $\|x\| \leq \|y\|$.

A complete normed vector lattice is called a Banach lattice.

Theorem 22. Let (E,C,u) be an ordered l.c. space such that (E,C) is a vector lattice. The following statements are equivalent:

(a) (E,C,u) is an l.c. vector lattice.

(b) For any nets $\{x_\alpha : \alpha \in A\}$ and $\{y_\alpha : \alpha \in A\}$ in E, if $\{y_\alpha : \alpha \in A\}$ converges to 0 for u and $|x_\alpha| \leq |y_\alpha|$ for all α, then $\{x_\alpha : \alpha \in A\}$ converges to 0 for u.

(c) There is a family $\{p_\beta : \beta \in \mathcal{B}\}$ of seminorms on E generating u such that $|x| \leq |y|$ implies $p_\beta(x) \leq p_\beta(y)$ for all $\beta \in \mathcal{B}$.

Remark 5. Each p_β of Theorem 22 (c) is called a lattice seminorm.

Proposition 38. Let (E,C,u) be an l.c. vector lattice. Then the completion $(\tilde{E},\tilde{u})$ of (E,u) is an l.c. vector lattice for the order determined by the closure $\tilde{C}$ of C in $(\tilde{E},\tilde{u})$.

Definition 29. Let (E,C) be an ordered vector space. The finest l.c. topology u for which every order-bounded set is u-bounded is called the order-bound topology.

We write u_b to denote the order-bound topology on (E,C).

Proposition 39. Let (E,C) be an ordered vector space. A neighbourhood basis of 0 for u_b is given by the class of all convex, circled and order-bornivorous sets.

Corollary 12. Let (E,C) be an ordered vector space. u_b is finer than any l.c. topology on E for which C is normal.

Proposition 40. Let (E,C,u) be a bornological l.c. vector lattice such that C is sequentially complete. Then $u = u_b$.

Corollary 13. Let (E,C,u) be a complete metrizable l.c. vector lattice. Then $u = u_b$.

Proposition 41. Let (E,C,u) be an l.c. vector lattice. Then $(E',C',\beta(E',E))$ is an l.c. vector lattice.

§12. The topology $0(F,E)$.

Suppose that $\langle E,F\rangle$ is a dual pair, C a cone in E and K the dual cone of C in F. Let $\mathcal{G}_0$ be the class of all order bounded subsets of E. From the general theory of $\mathfrak{S}$-topologies, it follows that the $\mathfrak{S}_0$-topology on F is compatible with the vector structure of F iff each element of F, considered as a linear functional on E, is order bounded i.e. $F \subset E^b$. If this condition is satisfied, then the $\mathfrak{S}_0$-topology on F is denoted by $0(F,E)$. If the cone C in E is generating, the family $\mathfrak{S}_0$ is directed ($\leq$) by inclusion so that the class $\{S^\circ: S \in \mathfrak{S}_0\}$ is a neighbourhood basis at 0 for the $\mathfrak{S}_0$-topology. Moreover the family $\{[-x,x]: x \in C\}$ is a fundamental system for $\mathfrak{S}_0$. The topology $0(F,E)$ is not in general compatible with the duality between E,F. (See [41], page 127 for a counter-example).

Proposition 42. If (E,C) is an ordered vector space, $F \subset E^b$, and E a full subset of the algebraic dual F^* of F, then $0(F,E)$ is consistent with the duality between E and F.

Corollary 14. If (E,C,u) is an ordered l.c. space with the closed and generating cone C and if E' is a full subspace of E^*, then $0(E,E')$ is consistent with the dual pair $\langle E,E'\rangle$.

Corollary 15. If (E,C,u) is an l.c. vector lattice, then $0(E,E')$ is consistent with $\langle E,E'\rangle$.

Remark 6. If (E,C,u) is an l.c. vector lattice, then we write $\sigma_s(E,E')$ for $0(E,E')$.

Proposition 43. If $\langle E,F \rangle$ is a dual pair, E a vector lattice and F a lattice ideal in E^b, then the band in E^b generated by F coincides with the $0(F,E)$-completion of F.

For more information about the $0(F,E)$ topology, we refer the reader to Peressini [41]. Also Peressini and Sherbert [42] initiated the study of ordered topological tensor products. The interested reader may consult [42] for more information about this topic.

Let (E_1,C_1) and (E_2,C_2) be ordered vector spaces. A wedge C in the tensor product $E_1 \otimes E_2$ is said to be compatible with E_1 and E_2 if $x \otimes y \in C$ whenever $x \in C_1$ and $y \in C_2$. The smallest compatible wedge in $E_1 \otimes E_2$ is called the projective wedge C_p defined by $C_p = \{ \sum_{n=1}^{k} x_n \otimes y_n : x_n \in C_1, y_n \in C_2 \}$.

Proposition 44. Let (E_1,C_1) and (E_2,C_2) be ordered vector spaces. If C_1 and C_2 are generating and if C is a compatible wedge in $E_1 \otimes E_2$, then C is generating.

Proposition 45. Let (E_1,C_1) and (E_2,C_2) be ordered vector spaces. Then each of the following conditions implies that the projective wedge C_p in $E_1 \otimes E_2$ is a cone:

(a) The cone C_1^* (or C_2^*) of all positive linear functionals on E_1 (or E_2) is total in E_1^* (or E_2^*).

(b) There is a strictly positive linear functional on E_1 (or E_2).

CHAPTER II

SOME CLASSES OF LOCALLY CONVEX SPACES

In this chapter, we recall some well-known classes of locally convex spaces and their properties with a view to generalizing them in subsequent chapters. Specifically, we give the definitions of barrelled, quasibarrelled, bornological, distinguished and DF-spaces, which have been introduced in the literature for the purpose of carrying over many theorems known to be true for Banach and Hilbert spaces.

§1. Barrelled and Quasibarrelled spaces.

The concepts of barrelled and quasibarrelled spaces as well as some of their properties can be found in Bourbaki [4]. A quasibarrelled space is a weaker form of a barrelled space. Barrelled spaces share some of the important properties of Banach and Fréchet spaces, but are neither necessarily complete nor metrizable ([4], Chapter 3, §1, Ex. 3,6). These spaces are successfully used in generalizing two of the most powerful theorems of functional analysis, namely the Banach-Steinhaus and the closed graph theorems.

Definition 1. Let (E,u) be an l.c. space.

(a) A subset B of E is said to be bornivorous if B absorbs

all u-bounded subsets of E.

(b) A closed, circled, convex and absorbing subset of E is said to be a barrel.

Remark 1. Every closed, circled and convex neighbourhood of 0 in (E,u) is a barrel, but not conversely.

Definition 2. An l.c. space (E,u) is called a barrelled (quasibarrelled) space if each barrel (respectively, bornivorous barrel) in (E,u) is a neighbourhood of 0.

Clearly each barrelled space is quasibarrelled. Examples will be given later to show that a quasibarrelled space need not be barrelled.

Proposition 1. Every Baire l.c. space is barrelled.

Proof: Let (E,u) be a Baire l.c. space and B a barrel in E. Since B is circled and absorbing, we have $E = \bigcup_{n=1}^{\infty} nB$. Since (E,u) is a Baire space, there exists n_0 such that the closed set n_0B has an interior point. Hence B itself has an interior point, say x_0. Since B is circled and convex, it follows that $0 = \frac{1}{2}x_0 + \frac{1}{2}(-x_0)$ is an interior point of B, and hence B is a neighbourhood of 0. This completes the proof.

Corollary 1. Every Fréchet space (in particular, Banach space) is a barrelled space.

Theorem 1. Let (E,u) be an l.c. space. Then the following statements are equivalent:

(a) (E,u) is barrelled.

(b) Any l.c. topology v on E with a base of neighbourhoods of 0 consisting of u-closed sets is weaker than u.

(c) Each $\sigma(E',E)$-bounded subset of E' is equicontinuous.

(d) Each lower semi-continuous seminorm on E is continuous.

Proof: (a) $\Rightarrow$ (b): Let V be a neighbourhood of 0 in (E,v). There is a neighbourhood W of 0 in (E,v) such that W is u-closed and $W \subset V$. Clearly there is a circled convex neighbourhood U of 0 in (E,v) such that $U \subset W \subset V$. But the u-closure of U is a barrel in (E,u) and hence a neighbourhood of 0 in (E,u). This implies that V is a neighbourhood of 0 in (E,u). Hence v is weaker than u.

(b) $\Rightarrow$ (a): Let B be a barrel in (E,u). Then the set of all barrels forms a base of neighbourhoods of 0 for an l.c. topology v which satisfies (b). Hence B is a neighbourhood of 0 in (E,u).

(a) $\Rightarrow$ (c): Let H be a $\sigma(E',E)$-bounded subset of E'. Then H^{o} is a closed, circled, convex and absorbing subset of E and hence a neighbourhood of 0. But then H^{oo} is equicontinuous (Chapter 1, Proposition 20). Hence H is equicontinuous, because $H \subseteq H^{oo}$.

(c) $\Rightarrow$ (a): Let B be a barrel in (E,u). Then $B = B^{oo}$ and so B^{o} is $\sigma(E',E)$-bounded (Chapter 1, Proposition 17(ii)). Hence by assumption, B^{o} is equicontinuous which implies that B is a neighbourhood of 0 in E.

(a) $\Leftrightarrow$ (d): This follows easily if we note that the set $V = \{x \in E: p(x) \leq 1\}$ is a barrel iff p is a lower semicontinuous seminorm.

Corollary 2. An l.c. space (E,u) is barrelled iff $u = \beta(E,E')$.

Corollary 3. Let (E,u) be a barrelled space and H a subset of E'. The following statements are equivalent:

(a) H is equicontinuous.

(b) H is relatively $\sigma(E',E)$-compact.

(c) H is $\beta(E',E)$-bounded.

(d) H is $\sigma(E',E)$-bounded.

Proof: (d) $\Leftrightarrow$ (a): This has been shown in Theorem 1.

(a) $\Rightarrow$ (b): This follows from Alaoglu-Bourbaki Theorem (Chapter I, Thorem 9).

(b) $\Rightarrow$ (c): This follows from (Chapter I, Proposition 23(à)).

(c) $\Rightarrow$ (d): This is obvious.

The following proposition is due to Saxon [46]. It can also be deduced from (Chapter 6, Corollary 2(i)). It gives a sufficient condition for a metric l.c. space to be barrelled.

Proposition 2. Let E be a metrizable l.c. space such that E' is $\sigma(E',E)$-sequentially complete. Then E is barrelled.

Proof: Suppose that $V_1 \supset V_2 \supset \dots$ is a countable base of neighbourhoods of 0 in E. Let H be a $\sigma(E',E)$-bounded subset of E'. We need to show that H is equicontinuous. Suppose not; then H is not uniformly bounded on V_n $(n \geq 1)$ and we claim the existence of a subsequence $\{V_{n_k}\}$ of $\{V_n\}$, a sequence $\{f_k\} \subset H$, and sequences $\{M_k\}$ and $\{x_k\}$ satisfying

(a) $M_k \geq 1, \quad k \geq 1$

(b) $x_k \in V_{n_k}, \quad k \geq 1$

(c) $|f_p(x_k)| \leq 1, \quad p < k, \ p,k \geq 1$

(d) $|f_k(x_k)| \geq k2^k M_k, \ k \geq 1$

(e) $|f_p(x_k)| \leq M_p$ for $p > k, \quad p,k \geq 1$.

Since H is not uniformly bounded, choose $M_1 = 1$, $n_1 = 1$, $f_1 \in H$ and $x_1 \in V_{n_1}$ such that $|f_1(x_1)| \geq 2M_1$. We observe that (a)-(e) are now satisfied for $p,k \leq 1$. Suppose $n_1 < \ldots < n_r$; $f_1,\ldots,f_r \in H$; $M_1,\ldots,M_r$ and $x_1,\ldots,x_r$ have been so chosen that (a)-(e) are satisfied whenever $p,k \leq r$. Then choose $n_{r+1} > n_r$ such that $|f_p(x)| \leq 1$ for all $x \in V_{n_{r+1}}$ and $p = 1,2,\ldots r$. Since H is pointwise bounded on E, there exists $M_{r+1} \geq 1$ such that $|f(x_k)| \leq M_{r+1}$ for all $f \in H$ and $k = 1,\ldots,r$. But H is not uniformly bounded on $V_{n_{r+1}}$, choose $f_{r+1} \in H$ and $x_{r+1} \in V_{n_{r+1}}$ such that

$$|f_{r+1}(x_{r+1})| \geq (r+1)2^{r+1}M_{r+1}.$$

Now (a)-(e) are satisfied for $p,k \leq r+1$, and the claim is established. Since $M_k \geq 1$, $k \geq 1$ and $\{f_k\} \subset H$ is $\sigma(E',E)$-bounded, we clearly have (absolute) convergence of the scalar series $\sum_{k=1}^{\infty} (2^k M_k)^{-1} f_k(x)$ for each $x \in E$. Since E' is $\sigma(E',E)$-sequentially complete, the linear functional f defined by

$$f(x) = \sum_{p=1}^{\infty} (2^p M_p)^{-1} f_p(x), \quad x \in E,$$

is continuous. But for $k \geq 1$,

$$|f(x_k)| \geq (2^k M_k)^{-1} |f_k(x_k)| - \sum_{p \neq k} (2^p M_p)^{-1} |f_p(x_k)|$$

$$\geq (2^k M_k)^{-1} k 2^k M_k - \sum_{p=1}^{k-1} (2^p M_p)^{-1} - \sum_{p=k+1}^{\infty} (2^p M_p)^{-1} M_p$$

$$\geq k - \sum_{p=1}^{\infty} (2^{-p}) = k-1.$$

Thus, $|f(x_k)| \to \infty$ which is a contradiction because f is continuous and $x_k \to 0$. Hence H is equicontinuous which completes the proof.

Similarly as for barrelled spaces, we give characterizations for quasibarrelled spaces.

Theorem 2. Let (E,u) be an l.c. space. Then the following statements are equivalent:

(a) (E,u) is quasibarrelled.

(b) Each $\beta(E',E)$-bounded subset of E' is equicontinuous.

(c) Each bounded lower semicontinuous seminorm p on E is continuous.

Proof: (a) $\Longrightarrow$(b): Let H be a $\beta(E',E)$-bounded subset of E'. Then H^o is a $\sigma(E,E')$-closed (and u-closed), circled and convex subset of E. Furthermore, H^o is bornivorous; for, given an u-bounded set A in E, there exists a $\lambda > 0$ such that $H \subset \lambda A^o$ and so $A \subset \lambda H^o$. Hence H^o is a neighbourhood of 0 in E from which it follows that H is equicontinuous.

(b) $\Longrightarrow$(a): Let B be a bornivorous barrel in (E,u). Then B^o is $\beta(E',E)$-bounded; for, given a bounded set A in (E,u), there exists a $\lambda > 0$ such that $A \subset \lambda B$ and so $B^o \subset \lambda A^o$. Hence B^o is equicontinuous from which it easily follows that $B^{oo} = B$ is a neighbourhood of 0.

(a) $\Leftrightarrow$ (c): This follows easily if we note that a set $V = (x \in E: p(x) \leq 1)$ is a bornivorous barrel iff p is a bounded lower semicontinuous seminorm.

<u>Proposition 3</u>. Let (E,u) be a quasibarrelled space. Then $u = \tau(E,E')$.

<u>Proof</u>: It is sufficient to show that the collection of circled, convex, $\sigma(E',E)$-closed and equicontinuous subsets of E' is identical with that of circled, convex and $\sigma(E',E)$-compact subsets of E'. By Alaoglu-Bourbaki theorem (Chapter I, Theorem 9), each $\sigma(E',E)$-closed equicontinuous subset of E' is $\sigma(E',E)$-compact. On the other hand, a convex and $\sigma(E',E)$-compact subset of E' is $\sigma(E',E)$=closed; and also $\beta(E',E)$-bounded (Chapter I, Proposition 23(b)) and therefore equicontinuous because (E,u) is quasibarrelled.

<u>Corollary 4</u>. Let (E,u) be a barrelled space. Then $u = \tau(E,E')$.

Let (E,u) be an l.c. space, E' its dual and E" its bidual. We write $\varepsilon(E'',E')$ to denote the topology of uniform convergence on all equicontinuous subsets of E'.

The canonical embedding $E \to E''$ is an algebraic isomorphism of E into E" given by $x \to f_x$ in which $f_x(x') = \langle x,x' \rangle$.

<u>Proposition 4</u>. Let (E,u) be a quasibarrelled space. Then $\beta(E'',E') = \varepsilon(E'',E')$ on E" (equivalently, the canonical embedding of E into E" is a topological isomorphism).

Proof: $\beta(E'',E') = \varepsilon(E'',E')$ means that every $\beta(E',E)$-bounded subset of E' is equicontinuous. By Theorem 2, this is the case iff E is quasibarrelled.

Theorem 3. An l.c. space (E,u) is reflexive iff it is semi-reflexive and quasibarrelled.

Proof: This follows easily from the definitions concerned and Proposition 4.

Corollary 5. Any reflexive space is barrelled.

Proof: Clear.

The following proposition shows when a quasibarrelled space is barrelled.

Proposition 5. A sequentially complete quasibarrelled space (E,u) is barrelled.

Proof: Let H be a $\sigma(E',E)$-bounded subset of E'. Since (E,u) is sequentially complete, H is $\beta(E',E)$-bounded (Chapter 1, Proposition 24) and hence, by Theorem 2, H is equicontinuous. This implies that (E,u) is barrelled by Theorem 1.

Corollary 6. A quasi-complete (in particular, complete) quasi-barrelled space is barrelled.

Lemma 1. Each barrel B in an l.c. space (E,u) absorbs every circled, convex, complete and bounded subset of (E,u).

Proof: Let M be a circled, convex, complete and bounded subset of (E,u). We can assume, without loss of generality, that M generates E because otherwise consider the subspace generated by M. Thus M is absorbing. Let p be the seminorm determined by M and $x \neq 0$. There exists a neighbourhood V of 0 such that $x \notin V$. Clearly there exists $\lambda > 0$ such that $\lambda M \subset V$. Therefore $p(x) \neq 0$ and so p is a norm on E. Let v be the norm topology on E defined by p. Since M is u-bounded it follows that $u \subset v$. Let $\{x_n\}$ be a Cauchy sequence in (E,v). Then for a given $\varepsilon > 0$, there exists $n_0 = n_0(\varepsilon)$ such that $p(x_{n+k}-x_n) < \varepsilon$ for all $n \geq n_0$, $k \geq 1$. In other words, $x_{n+k} \in x_n + \varepsilon M$ for all $n \geq n_0$, $k \geq 1$. Fixing $N \geqslant n_0$, $\{x_{n+k}\}_{k \geq 1}$ is a Cauchy sequence in (E,v) and therefore in (E,u). Since M is u-complete, there exists $x \in x_n + \varepsilon M$ and $x_{n+k} \to x$. Hence $p(x-x_n) < \varepsilon$ for all $n \geq n_0$. In other words, (E,v) is a Banach space and so barrelled. Clearly B is a barrel in (E,v) and hence a neighbourhood of 0 in (E,v). It now follows that B absorbs M.

<u>Proposition 6.</u> The completion $(\tilde{E},\tilde{u})$ of a quasibarrelled space (E,u) is barrelled.

<u>Proof</u>: Let B be a barrel in $(\tilde{E},\tilde{u})$. Clearly $B \cap E$ is a barrel in (E,u). Let M be a bounded subset of E. Then the closure $\overline{h(M)}$ of the circled convex hull h(M) (taken in $\tilde{E}$) of M is a circled, convex, complete and bounded subset of E. Hence, by Lemma 1, B absorbs $\overline{h(M)}$ and so $B \cap E$ absorbs M. Thus $B \cap E$ is a neighbourhood of 0 in E. But then $\overline{B \cap E} = B$ is a neighbourhood of 0 in $(\tilde{E},\tilde{u})$ and this completes the proof.

Theorem 4. Let (E,u) be a barrelled (quasibarrelled) space and (F,v) any l.c. space. Then each pointwise bounded (strongly bounded) subset H of $\mathcal{L}(E,F)$ is equicontinuous.

Proof: Let V be a closed, circled and convex neighbourhood of 0 in F and $B = \bigcap_{f \in H} f^{-1}(V)$. Clearly B is closed, circled and convex; also it is absorbing (bornivorous) because H is pointwise bounded (strongly bounded). Thus B is a barrel (bornivorous barrel) in (E,u) and hence a neighbourhood of 0 in (E,u). This implies that H is equicontinuous.

A consequence of Theorem 4 is the following.

Corollary 7. (Banach-Steinhaus theorem) Let (E,u) be a barrelled space and (F,v) any l.c. space. Let $\mathfrak{F}$ be a pointwise bounded filter on $\mathcal{L}(E,F)$ which converges pointwise to a map f_0: $E \to F$. Then $f_0 \in \mathcal{L}(E,F)$ and $\mathfrak{F}$ converges to f_0 uniformly on precompact sets.

Proof: Since $\mathfrak{F}$ is a pointwise bounded filter or, in other words, $\mathfrak{F}$ contains a pointwise bounded set H, it follows that H is equicontinuous by Theorem 4. Hence the corollary follows from (Chapter 1, §6, Corollary 6).

As an application of Theorem 4, we have the following weak basis theorem which is due to Arsove and Edwards [2].

Theorem 5. Let (E,u) be a barrelled space. Then every $\sigma(E,E')$-Schauder basis in E is an u-Schauder basis.

Proof: Let $\{x_i, f_i\}$ be a $\sigma(E,E')$-Schauder basis in E. Then $\{x_i, f_i\}$ is a weak Markuschevich basis (Chapter 1, Theorem 15(i)). Thus $\{x_i\}$ is a total set in E. This implies that we can choose, for each $x \in E$, a sequence $\{y_n\} \subset E$ such that $y_n \to x$ and $y_n \in sp\{x_i : 1 \le i \le n\}$. Consider the sequence $\{T_n\}$ of continuous linear maps $T_n: E \to E$ defined by $T_n(x) = \sum_{i=1}^{n} f_i(x)x_i$, $x \in E$, $n \ge 1$. We observe that $\lim_n T_n(x) = x$ in the weak topology of E; hence $\{T_n\}$ is pointwise bounded. But then, by Theorem 4, $\{T_n\}$ is equicontinuous. Hence (taking limit in u-topology) we have:

$$\begin{aligned} x &= x + \lim_n T_n(x - y_n) \\ &= x + \lim_n \sum_{i=1}^{n} f_i(x - y_n)x_i \\ &= \lim_n y_n + \lim_n \left(\sum_{i=1}^{n} f_i(x)x_i - y_n\right) \\ &= \lim_n \sum_{i=1}^{n} f_i(x)x_i, \end{aligned}$$

and so $\{x_i, f_i\}$ is a Schauder basis for u.

As another application of the Banach-Steinhaus theorem, we have the following result, called the isomorphism theorem, which is due to Jones and Rutherford [24].

Theorem 6. Let (E,u) and (F,v) be barrelled spaces and $\{x_n, f_n\}$ and $\{y_n, g_n\}$ Schauder bases in E and F respectively. Then $\{x_n, f_n\}$ is similar (cf. Def. 23, §9, Chapter 1) to $\{y_n, g_n\}$ iff there exists a topological isomorphism $T: E \to F$ such that

$Tx_n = y_n$ for all $n \geq 1$.

Proof: Assume that there is an isomorphism $T: E \to F$ such that $Tx_n = y_n$ for all $n \geq 1$. Then, clearly $\sum_{n=1}^{\infty} a_n x_n$ converges iff $T(\sum_{n=1}^{\infty} a_n x_n) = \sum_{n=1}^{\infty} a_n Tx_n = \sum_{n=1}^{\infty} a_n y_n$ converges. Hence we get similarity. Conversely, assume that $\{x_n, f_n\}$ and $\{y_n, g_n\}$ are similar Schauder bases in E and F respectively. For $x \in E$, $x = \sum_{n=1}^{\infty} f_n(x)x_n$. Consider the sequence $\{T_m\}$ of continuous linear maps T_m: $E \to F$ defined by $T_m(x) = \sum_{n=1}^{m} f_n(x)y_n$, $m \geq 1$. Define T: $E \to F$ by $T(x) = \sum_{n=1}^{\infty} f_n(x)y_n$. T is well-defined, because by similarity of the basis, $\sum_{n=1}^{\infty} f_n(x)y_n$ is convergent. T is one to one: $Tx = 0$ implies $\sum_{n=1}^{\infty} f_n(x)y_n = 0$, and (since $\{y_n, g_n\}$ is a Schauder basis) $f_n(x) = 0$ for all $n \geq 1$. Hence $x = 0$ because $\{f_n\}$ is separating. T is onto: For $y \in F$, $y = \sum_{n=1}^{\infty} b_n y_n$, and by similarity,

$$\sum_{n=1}^{\infty} b_n x_n = \sum_{n=1}^{\infty} f_n(x)x_n \text{ converges in } (E,u) \text{ to some } x \in E,$$

hence $Tx = y$. Clearly $T_m(x) \to T(x)$ for each $x \in E$, and so $\{T_m\}$ is pointwise bounded. Hence, by Corollary 7, T is continuous and linear. By symmetry the same is true of T^{-1} and hence T is the desired isomorphism of E onto F.

The following example [2] shows that the isomorphism theorem does not hold for generalized bases even in complete barrelled spaces.

Example 1. Let (E,u) be an infinite dimensional Banach space with $\{x_\alpha, f_\alpha\}$ a generalized basis and let (F,v) be E with the strongest l.c. topology. Then (E,u) and (F,v) are complete barrelled spaces. Also $\{x_\alpha, f_\alpha\}$ is a generalized basis in (F,v) and is trivially similar. However (E,u) and (F,v) are not isomorphic because v is strictly finer than u.

In view of Example 1, the following result which is due independently to Johnson and Dyer [23], and Bozel and Husain [5] is in a sense the most general one.

Theorem 7. Let (E,u) and (F,v) be B-complete barrelled spaces and $(x_\alpha)_{\alpha\varepsilon A}$ a generalized basis in (E,u). If $T: E \to F$ is an isomorphism such that $Tx_\alpha = y_\alpha$ for all $\alpha \varepsilon A$, then $(y_\alpha)_{\alpha\varepsilon A}$ is a generalized basis in (F,v) and (x_α) is similar to (y_α). Conversely, if $(y_\alpha)_{\alpha\varepsilon A}$ is a generalized basis in (F,v) and (x_α) is similar to (y_α), then there exists an isomorphism $T: E \to F$ such that $Tx_\alpha = y_\alpha$ for all $\alpha \varepsilon A$.

Proof: Let $\{x_\alpha, f_\alpha\}_{\alpha\varepsilon A}$ be a generalized basis in (E,u) and $T: E \to F$ an isomorphism such that $Tx_\alpha = y_\alpha$ for all $\alpha \varepsilon A$. Define $(g_\alpha)_{\alpha\varepsilon A}$ by $g_\alpha = f_\alpha \circ T^{-1}$. Then the g_α's are continuous, linear and

$$g_\alpha(Tx_\beta) = (f_\alpha \circ T^{-1})(Tx_\beta) = f_\alpha(x_\beta) = \delta_{\alpha\beta}.$$

Hence $\{Tx_\alpha, g_\alpha\}_{\alpha\varepsilon A}$ is biorthogonal. Furthermore, if $g_\alpha(y) = 0$ for all $\alpha \varepsilon A$, then $(f_\alpha \circ T^{-1})(y) = 0$ for all $\alpha \varepsilon A$ and hence $T^{-1}(y) = 0$. This implies that $y = 0$ because T is an isomorphism. Since

$$\Psi(y) = (g_\alpha(y))_{\alpha\varepsilon A} = (g_\alpha(Tx))_{\alpha\varepsilon A} = (f_\alpha(x))_{\alpha\varepsilon A} = \Phi(x),$$

Ψ and Φ being the corresponding coefficient maps, it follows that $\{x_\alpha, f_\alpha\}$ is similar to $\{y_\alpha, g_\alpha\}$. Conversely, assume that $\{y_\alpha\}_{\alpha\varepsilon A}$ is a generalized basis in (F,v) and that $\{x_\alpha\}$ is similar to (y_α). Then $\Phi(E) = \Psi(F) = G$, say. Clearly G is a subspace of $\mathbb{R}^A$ and is algebraically isomorphic to E and F respectively. Hence we can induce the topologies u and v on G as follows: $A \subset G$ is open iff $\Phi^{-1}(A)$ (or $\Psi^{-1}(A)$) is open in (E,u) (respectively, in (F,v)). Hence (G,u) and (G,v) are B-complete barrelled spaces. Clearly $T = \Psi^{-1} \circ \Phi$ is a linear one-to-one map of E onto F. Define $(h_\alpha)_{\alpha\varepsilon A}$ by $h_\alpha(z) = f_\alpha(z) = g_\alpha(y)$, $z = \Phi(x) = \Psi(y)$. Then $\{h_\alpha\}$ is a separating family of continuous linear functionals and hence the identity $i: (G,u) \to (G,v)$ has closed graph. Hence, by the closed graph theorem (Theorem 8'(ii), §5, Ch.I), u is finer than v which implies that $u = v$ by the open mapping theorem (Theorem 8' (i), §5, Ch.I). Thus T is the desired isomorphism. That $Tx_\alpha = y_\alpha$ for all $\alpha \varepsilon A$ follows from similarity.

As an application of Theorem 7, we have the following

<u>Corollary 8</u>. Let (E,u) and (F,v) be B-complete barrelled spaces and $\{x_\alpha, f_\alpha\}$ and $\{y_\alpha, g_\alpha\}$ as Markuschevich bases in E and F, respectively. Let $A_0 \subseteq A$ be such that $A \backslash A_0$ is finite. Let $E_1 = [x_\alpha]_{\alpha\varepsilon A_0}$ and $F_1 = [y_\alpha]_{\alpha\varepsilon A_0}$ be the closed linear spans of $\{x_\alpha\}_{\alpha\varepsilon A_0}$ and $\{y_\alpha\}_{\alpha\varepsilon A_0}$ with the relative topologies of E and F respectively. If T is an isomorphism of E_1 onto F_1 such that

$Tx_\alpha = y_\alpha$, $\alpha \in A_0$, then T can be extended to an isomorphism of E onto F such that $Tx_\alpha = y_\alpha$ for all $\alpha \in A$.

<u>Proof</u>: That $\{x_\alpha, f_\alpha\}_{\alpha\in A_0}$ and $\{y_\alpha, g_\alpha\}_{\alpha\in A_0}$ are Markuschevich bases in E_1 and F_1 respectively follows from the definition of Markuschevich bases. Since T is an isomorphism and $Tx_\alpha = y_\alpha$, $\alpha \in A_0$, the similarity of bases follows. Since the cardinality of $A \backslash A_0$ is finite, $[x_\alpha]_{\alpha\in A\backslash A_0}$ is isomorphic to $[y_\alpha]_{\alpha\in A\backslash A_0}$ and x_α goes to y_α. By hypothesis, E_1 is isomorphic to F_1 and x_α goes to y_α. Clearly $E = E_1 \oplus [x_\alpha]_{\alpha\in A\backslash A_0}$ and $F = F_1 \oplus [y_\alpha]_{\alpha\in A\backslash A_0}$. Hence $\{x_\alpha, f_\alpha\}_{\alpha\in A}$ is similar to $\{y_\alpha, g_\alpha\}_{\alpha\in A}$. Now Theorem 7 can be applied to get the desired isomorphism.

As another application of Theorem 7, we have the following:

<u>Corollary 9</u>. Let (E,u) and (F,v) be B-complete barrelled spaces and $\{x_\alpha, f_\alpha\}_{\alpha\in A}$ and $\{y_\alpha, g_\alpha\}_{\alpha\in A}$ similar biorthogonal systems for E and F respectively. Then there exists an isomorphism T of $E/\mathrm{Ker}(\Phi)$ onto $F/\mathrm{Ker}(\Psi)$ such that $T(\phi(x_\alpha)) = \phi'(y_\alpha)$ for all $\alpha \in A$ where ϕ and ϕ' are quotient maps and Φ and Ψ the respective coefficient maps.

<u>Proof</u>: Define $\tilde{f}_\alpha$ and $\tilde{g}_\alpha$ by

$$\tilde{f}_\alpha(\phi(x)) = f_\alpha(x), \quad \tilde{g}_\alpha(\phi'(y)) = g_\alpha(y).$$

Then $\{\phi(x_\alpha), \tilde{f}_\alpha\}_{\alpha\in A}$ and $\{\phi'(y_\alpha), \tilde{g}_\alpha\}_{\alpha\in A}$ are generalized bases in $E/\mathrm{Ker}(\Phi)$ and $F/\mathrm{Ker}(\Psi)$, respectively. Similarity of these bases follows easily because $\{x_\alpha, f_\alpha\}$ is similar to $\{y_\alpha, g_\alpha\}$. Now we topologize $\phi(E)$ and $\psi(F)$ with the topologies induced from $E/\mathrm{Ker}(\Phi)$ and $F/\mathrm{Ker}(\Psi)$ by $\tilde{\Phi}$ and $\tilde{\Psi}$ respectively. Hence $\Phi = \tilde{\Phi} \circ \phi$, $\Psi = \tilde{\Psi} \circ \phi'$ are continuous and so $\mathrm{Ker}(\Phi)$ and $\mathrm{Ker}(\Psi)$ are closed subspaces of E and F respectively. Since $E/\mathrm{Ker}(\Phi)$ and $F/\mathrm{Ker}(\Psi)$ are B-complete barrelled spaces, Theorem 7 applies.

We now show that barrelled spaces are well suited for carrying over the celebrated theorems of functional analysis, namely the closed graph and the open mapping theorems.

Lemma 2. Let (E,u) be a barrelled space and (F,v) any l.c. space.

(a) Any linear map f of E into F is almost continuous.

(b) Any linear map g of F onto E is almost open.

Proof: (a) Let V be a closed, circled and convex neighbourhood of 0 in (F,v). Then $\overline{f^{-1}(V)}$ is a closed, circled and convex set in (E,u); also it is absorbing because V is absorbing and f linear. Hence $\overline{f^{-1}(V)}$ is a barrel in (E,u) and so a neighbourhood of 0 in (E,u).

(b) Let W be a closed, circled and convex neighbourhood of 0 in (F,v). Then $\overline{g(W)}$ is a barrel in (E,u) and hence a neighbourhood of 0 in (E,u). This completes the proof.

Theorem 8. Let (E,u) be a barrelled space and (F,v) a Fréchet space.

(a) A linear map f of E into F with closed graph is continuous.

(b) A continuous linear map of F onto E is open.

Proof: (a) By Lemma 2, Part (a), f is almost continuous. Hence f is continuous by (Chapter I, Theorem 7(a)).

(b) By Lemma 2, Part (b), g is almost open. It now follows from (Chapter I, Theorem 7(b)) that g is open.

The proofs of the following permanence properties of barrelled and quasibarrelled spaces are omitted, since they are similar to the ones adopted in Chapter 5, for more general situations.

(1) The inductive limit of any family of barrelled (quasibarrelled) spaces is barrelled (quasibarrelled).

(2) Each quotient space of a barrelled (quasibarrelled) space is barrelled (quasibarrelled).

(3) The topological direct sum of barrelled (quasibarrelled) spaces is barrelled (quasibarrelled).

(4) The product of any family of barrelled (quasibarrelled) spaces is barrelled (quasibarrelled).

(5) Let (E,u) and (F,v) be any l.c. spaces and f a linear, continuous and almost open map of (E,u) into (F,v). If (E,u) is barrelled, so is (F,v).

A closed subspace of a barrelled (or quasibarrelled) space need not be barrelled (or quasibarrelled) [31], page 434. See also Chapter 5, Example 5). But it is a well known fact that a finite codimensional subspace of a barrelled space is barrelled ([31], page 369). Valdivia [52] and Saxon and Levin [45] have shown, however, that every infinite countable codimensional subspace of a barrelled space is barrelled. We need the following lemmas to prove that result.

Lemma 3. Let (E,u) be a barrelled space and $\{A_n; n=1,2,..\}$ an increasing sequence of circled convex subsets such that $A = \bigcup_{n\geq 1} A_n$ is absorbing. Let $\mathfrak{F}$ be a Cauchy filter in A and $\mathfrak{G}$ the filter whose base is $\{M+U\}$ where M varies in $\mathfrak{F}$ and U over the filter of the neighbourhoods of 0 in E. Then there exists an integer $n_0 > 0$ such that $\mathfrak{G}$ induces a filter on $2A_{n_0}$.

Proof: Suppose that the lemma is false. Then there exists a decreasing sequence $\{V_n; n=1,2,...\}$ of circled and convex neighbourhoods of 0 and a sequence $\{M_n\}$ of elements of $\mathfrak{F}$ such that $M_n - M_n \subset V_n$ and $(M_n+V_n) \cap 2A_n = \emptyset$, $n = 1,2,...$. If W is the convex hull of $\bigcup_{n\geq 1} \tfrac{1}{2}(V_n \cap A_n)$, then W is absorbing because A is absorbing. Therefore there exists an integer n_1 such that $\lambda x \in \tfrac{1}{2}A_{n_1}$. Since there exists an $M \in (0,1)$ such that

$$M\lambda x \in \tfrac{1}{2}V_{n_1}, \; M\lambda x \in \tfrac{1}{2}(V_{n_1} \cap A_{n_1}) \subset W.$$

If W_n is the convex hull of $\tfrac{1}{2}(V_1 \cap A_1) \cup \tfrac{1}{2}(V_2 \cap A_2) \cup \cdots \cup \tfrac{1}{2}(V_{n-1} \cap A_{n-1}) \cup \tfrac{1}{2}V_n$, then W_n is a neighbourhood of 0 and clearly $W \subset W_n$. If $\tilde{W}$ and $\tilde{W}_n$ are the closures of W and W_n respectively, then $\tfrac{1}{2}\tilde{W} \subset \tfrac{1}{2}\tilde{W}_n \subset W_n$. Since W is circled, convex and absorbing, W is a barrel in E and hence a neighbourhood of 0. Hence there exists an element $P \in \mathfrak{F}$ such that $P-P \subset \tfrac{1}{2}\tilde{W}$. Let $P_n \in \mathfrak{F}$ be such that $P_n - P_n \subset W_n$. We show that $(P_n+W_n) \cap A_n = \emptyset$.

If $x_0 \in P_n \cap M_n \in \mathfrak{J}$ and $y \in P_n$, then $y-x_0 \in P_n-P_n \subset W_n$ which implies that $y = x_0 + \sum_{p=1}^{n} a_p x_p$, $a_p \geq 0$, $p = 1,2,\ldots,n$, and $\sum_{p=1}^{n} a_p = 1$. Also $x_p \in \frac{1}{2}(V_p \cap A_p)$, $p = 1,2,\ldots n-1$, $x_n \in \frac{1}{2}V_n$.

If $z \in P_n + W_n$, we can write

$$z = x_0 + \sum_{p=1}^{n} a_p x_p + \sum_{p=1}^{n} b_p y_p,\ b_p \geq 0,\ p = 1,2,\ldots n,\ \sum_{p=1}^{n} b_p = 1,$$

$y_p \in \frac{1}{2}(V_p \cap A_p)$, $p = 1,2,\ldots,n-1$, $y_n \in \frac{1}{2}V_n$.

Clearly

$$\sum_{p=1}^{n-1} a_p x_p \in \tfrac{1}{2}A_{n-1},\quad \sum_{p=1}^{n-1} b_p y_p \in \tfrac{1}{2}A_{n-1}$$

and hence

$$\gamma = \sum_{p=1}^{n-1} a_p x_p + \sum_{p=1}^{n-1} b_p y_p \in \tfrac{1}{2}A_{n-1} + \tfrac{1}{2}A_{n-1} = A_{n-1} \subset A_n.$$

Furthermore, $a_n x_n \in \frac{1}{2}V_n$, $b_n y_n \in \frac{1}{2}V_n$ and therefore

$$a_n x_n + b_n y_n \in \tfrac{1}{2}V_n + \tfrac{1}{2}V_n = V_n.$$

Since $x_0 \in M_n$, we obtain

$$x_0 + a_n x_n + b_n y_n \in M_n + V_n$$

and hence

$$(*) \qquad x_0 + a_n x_n + b_n y_n \not\in 2A_n.$$

If $z \in A_n$, then

$$x_0 + a_n x_n + b_n y_n = z-\gamma \in A_n + A_n = 2A_n$$

which contradicts (*). Hence

$$(P_n + W_n) \cap A_n = \emptyset.$$

If $z_0 \in P$, there exists an integer $n_2 > 0$ such that $z_0 \in A_{n_2}$.

If $w \in P_{n_2}$, then $z_0 - w \notin W_{n_2}$. Since $\frac{1}{2}\tilde{W} \subset W_{n_2}$, we obtain $z_0 - w \notin \frac{1}{2}\tilde{W}$. Furthermore, $P-P \subset \frac{1}{2}\tilde{W}$ and $z_0 \in P$ and hence we deduce that $w \notin P$, i.e. $P \cap P_{n_2} = \emptyset$ which is not true. Hence the lemma is proved.

Lemma 4. Let $\{W_n;\ n \geq 1\}$ be an increasing sequence of circled and convex sets in a barrelled space (E,u) such that $E = \bigcup_{n=1}^{\infty} W_n$. If V is a circled, convex and absorbing set such that $V \cap W_n$ is closed in W_n, $n = 1,2,\ldots$, then V is a neighbourhood of 0 in E.

Proof: Put $A_n = V \cap W_n$, $n \geq 1$. Clearly, $\bigcup_{n\geq 1} A_n = V$. Let $x \in \tilde{V}$ (the completion of V). Clearly the sequence $\{A_n\}$ satisfies the conditions of Lemma 3. Hence if $\mathfrak{F}$ is a Cauchy filter in V, converging to x, then it is possible to find an integer $n_0 > 0$ such that the filter $\mathfrak{G}$ described in Lemma 3 induces a filter $\mathfrak{G}^1$ in $2A_{n_0}$. It is immediately clear that n_0 can be taken in such a way that $x \in W_{n_0}$. Since A_{n_0} is closed in W_{n_0}, $2A_{n_0}$ is closed in $2W_{n_0}$ and hence $x \in 2A_{n_0} \subset 2V$, i.e., $\tilde{V} \subset 2V$. Since $\tilde{V}$ is a barrel in E, $\tilde{V}$ and so $2V$ is a neighbourhood of 0 in E.

Corollary 10. Let $\{E_n;\ n \geq 1\}$ be an increasing sequence of subspaces of a barrelled space (E,u) such that $E = \bigcup_{n\geq 1} E_n$. If V is a set such that $V \cap E_n$, $n \geq 1$, is a barrel in E_n, then V is a neighbourhood of 0 in E.

The following lemma is due to De Wilde [6].

Lemma 5. Let (E,u) be an l.c. space and M a finite codimensional subspace of E. If B is a barrel in M, there exists a barrel B' in E such that $B' \cap M = B$.

Proof: Let $\overline{B}$ be the closure of B in E. Let M' be the linear hull of $\overline{B}$. If $M' = E$, then $\overline{B} \equiv B'$ is a barrel in E such that $B' \cap M = B$. If $M' \neq E$, let $\{e_k; 1 \leq k \leq m\}$ be a cobase of M'. The set $\overline{B} + A_m$, where A_m is the circled convex hull of $e_1,..,e_m$, is a barrel in E. Since $e_1,\ldots,e_m$ is a cobase of M', it follows that $(\overline{B} + A_m) \cap M' = \overline{B}$, and hence $(\overline{B} + A_m) \cap M = \overline{B} \cap M = B$.

Corollary 11. If E is a barrelled space and M a finite codimensional subspace of E, then M is barrelled.

Theorem 9. Let (E,u) be a barrelled space and E_1 an infinite countable codimensional subspace of E. Then E_1, with the induced topology, is barrelled.

Proof: Let $\{x_n; n \geq 1\}$ be an infinite countable set of vectors such that $E_1 \cup \{x_n; n \geq 1\}$ generates E. Let E_n denote the space generated by $E_1 \cup \{x_1,x_2,\ldots,x_{n-1}\}$, $n \geq 2$, and let V_1 be a barrel in E_1. Since E_1 is a finite codimensional subspace of E_2, there exists, by Lemma 5, a barrel V_2 in E_2 such that $V_2 \cap E_1 = V_1$. Proceeding inductively, if V_n is a barrel in E_n, we can find a barrel V_{n+1} in E_{n+1} such that $V_{n+1} \cap E_n = V_n$. If $V = \bigcup_{n\geq 1} V_n$, then $V \cap E_n = (\bigcup_{p\geq 1} V_p) \cap E_n = (\bigcup_{p\geq n} V_p) \cap E_n = V_n$, $n \geq 1$. Hence by Corollary 10, V_1 is a neighbourhood of 0 in E_1 which implies that E_1 is barrelled.

A simple adaptation of Lemma 3 and Proposition 11 of Chapter 5, enables us to obtain the following two results.

Lemma 6. Let E be a quasibarrelled space and M a closed subspace of E of countable codimension such that for each bounded subset B of E, M is of finite codimension in the span of $(M \cup B)$. Then M is quasibarrelled.

We write E^+ to denote the set of all sequentially continuous linear functionals on the l.c. space E and the elements of E^+ are given by the sequentially closed hyperplanes in E [56].

Proposition 7. Let E be a quasibarrelled space with $E' = E^+$. If M is a subspace of E such that M is of countable codimension in E and M is of finite codimension in span $(M \cup B)$ for every bounded set B, then M is quasibarrelled.

Proof: Easy.

Proposition 8. An arbitrary l.c. space E is a closed subspace of some barrelled space.

Proof: Let F be a barrelled space containing the completion of E and $\{e_\alpha\}$, a Hamel basis of the complement space F^c of F. For any index α, let F_α be the subspace of F generated by E and $\{e_\beta\colon \beta \neq \alpha\}$. Since F_α is of codimension 1, it is barrelled. The space E is canonically embedded in $\prod_\alpha F_\alpha$ which is barrelled. If some directed set $(x_\alpha)^\phi = (x_\alpha{}^\phi)$ in E converges to $(y_\alpha) \in \prod_\alpha F_\alpha$, then for any α, $(x_\alpha)^\phi$ converges to y_α. Clearly y_α is independent of α. Thus $y_\alpha = y \in \bigcap_\alpha F_\alpha = E$, which implies that E is closed in $\prod_\alpha F_\alpha$.

The following important theorem is due to Bourbaki [4]. For the proof, see (Chapter 7, Theorem 4).

<u>Theorem 10</u>. Let E be a metrizable barrelled space, F a metrizable topological space and G an l.c. space. Let H be a family of mappings of $E \times F$ into G such that

(i) for each fixed $y \in F$, $\{f(\cdot,y);\ f \in H\}$ is an equicontinuous subset of $\mathcal{L}(E,G)$, and

(ii) for each fixed $x \in E$, $\{f(x,\cdot);\ f \in H\}$ is an equicontinuous subset of $\mathcal{L}(F,G)$.

Then H is an equicontinuous family of maps of $E \times F$ into G.

<u>Corollary 12</u>. If E, F and G are as in Theorem 10, then every separately continuous bilinear map of $E \times F$ into G is continuous.

<u>Proposition 9</u>. The (Projective) tensor product of two metrizable barrelled spaces is barrelled.

<u>Proof</u>: Let V be a barrel in the projective tensor product $(E \otimes F, u_p)$ of metrizable barrelled spaces (E,u) and (F,v). Let τ be the l.c. topology on $E \otimes F$ with all the barrels as a base of neighbourhoods of 0. Since the canonical map $g: E \times F \to (E \otimes F, u_p)$ is continuous, each partial map $g_y: E \to (E \otimes F, u_p)$, $y \in F$, is continuous. Hence $g_y^{-1}(V)$ is a barrel in E and so a

neighbourhood of 0 in E. This implies that the map of g_y: $E \to (E \otimes F, \tau)$ is continuous. It follows easily from here that the map g: $E \times F \to (E \otimes F, \tau)$ is separately continuous and, by Corollary 12, it is continuous. Thus $\tau \subseteq u_p$ and so V is a neighbourhood of 0 in $(E \otimes F, u_p)$.

§2. Bornological spaces.

Definition 3. An l.c. space (E,u) is said to be bornological if each circled, convex and bornivorous subset of (E,u) is a neighbourhood of 0.

Proposition 10. Every metrizable l.c. space is bornological.

Proof: Let $\{V_n;\ n \geq 1\}$, $V_n \supset V_{n+1}$ for all $n \geq 1$ be a neighbourhood basis of 0 in (E,u). Let B be a circled, convex and bornivorous subset of (E,u). We claim that $V_n \subset n\,B$ for some $n \in \mathbb{N}$. Suppose not, then there exist elements $x_n \in V_n$ such that $x_n \notin n\,B$, $n \in \mathbb{N}$. But this is not true, because $\{x_n\}$ being a sequence converging to 0 is bounded and so it is absorbed by B. Thus $V_n \subset n\,B$ for some $n \in \mathbb{N}$ which shows that B is a neighbourhood of 0 in (E,u).

Proposition 11. Every bornological space is quasibarrelled.

Proof: This is obvious in view of the definitions concerned.

Corollary 13. Every sequentially complete bornological space is barrelled.

Proof: This follows from Propositions 5 and 11

<u>Corollary 14</u>. If (E,u) is a bornological space, then $u = \tau(E,E')$.

<u>Proof</u>: Combine Propositions 3 and 11.

<u>Corollary 15</u>. The completion of a bornological space is barrelled.

<u>Remark</u>. There exist barrelled spaces which are not bornological (Nachbin [38], Shirota [49]). The following example shows that a bornological space need not be barrelled.

<u>Example 2</u>. Let ϕ be the normed vector space of all sequences with only finitely many non-zero components and equipped with the supremum norm $\|.\|$. By Proposition 10, ϕ is a bornological space. But we show that it is not barrelled. For each $n \in \mathbb{N}$, let f_n be the continuous linear functional on ϕ defined by $f_n(x) = nx_n$. Then the set $\{f_n; n \geq 1\}$ is pointwise bounded. However, it can easily be shown that $\{f_n\}$ is not equicontinuous. Hence by Theorem 1, it follows that ϕ is not barrelled.

We now give an example, which is due to Mahowald and Gould [33], of a quasibarrelled space which is neither barrelled nor bornological.

<u>Example 3</u>. Let E be a barrelled space which is not bornological and F a bornological space which is not barrelled. Then $E \times F$ being the product of two quasibarrelled spaces, is quasibarrelled. However if A is a barrel in F which is not a neighbourhood of 0, then $E \times A$ will be a barrel in $E \times F$ but is not a neighbourhood of 0. This implies that $E \times F$ is not barrelled. Similarly if

B is a bornivorous circled convex set in E which is not a neighbourhood of 0, then $B \times F$ is a bornivorous circled convex set in $E \times F$, but is not a neighbourhood of 0. This shows that $E \times F$ is not bornological.

Theorem 11. An l.c. space (E,u) is bornological iff each seminorm p on E that is bounded is continuous.

Proof: Assume that (E,u) is bornological and p a bounded seminorm on E. Then the set $V = \{x \in E: p(x) \leq 1\}$ is clearly circled, convex and bornivorous in (E,u) and hence a neighbourhood of 0 in (E,u). This proves that p is continuous. For the converse, let V be a circled, convex and bornivorous subset of E. Then its gauge p is a bounded seminorm and hence continuous. This implies that V is a neighbourhood of 0 and the proof is complete.

We now characterize bornological spaces as those l.c. spaces (E,u) for which every bounded linear functional is continuous. For this we can assume that u is the Mackey topology, because we are concerned with a property which depends only upon the dual pair $\langle E', E \rangle$ and not upon u.

Theorem 12. Let (E,u) be an l.c. space. Every bounded linear functional on (E,u) is continuous iff $(E, \tau(E,E'))$ is bornological.

Proof: For the "only if" part, consider the l.c. topology u_0 on E with a base of neighbourhoods of 0 consisting of circled, convex and bornivorous sets. Then clearly u_0 is finer than $\tau(E,E')$.

This implies that $(E,\tau(E,E'))$ is bornological. For the "if" part, we note that if (E,u) is bornological, then $u = \tau(E,E')$ by Corollary 14. Let f be a bounded linear functional on E. Then the set $V = \{x \in E;\ |\langle x,f\rangle| \leq 1\}$ is circled, convex and bornivorous and hence a neighbourhood of 0 in (E,u). This proves that f is continuous.

<u>Theorem 13</u>. An l.c. space (E,u) is bornological iff each bounded linear map of (E,u) into any l.c. space (F,v) is continuous.

<u>Proof</u>: Assume that (E,u) is bornological and $f: E \to F$ a bounded linear map. If V is a circled convex neighbourhood of 0 in F, then it follows easily that $f^{-1}(V)$ is circled, convex and bornivorous in (E,u) and hence a neighbourhood of 0. This proves that f is continuous. The "only if" part follows from Theorem 12 if we note that the identity map is a bounded linear map of (E,u) onto $(E,\tau(E,E'))$ and so $u = \tau(E,E')$.

<u>Corollary 16.</u> Let (E,u) be a bornological space and f a linear map of (E,u) into an l.c. space (F,v). The following statements are equivalent:

(a) f is continuous.

(b) For each null sequence $\{x_n\} \subset E$, $\{f(x_n)\}$ is a null sequence.

(c) f is bounded.

<u>Proof</u>: (a) $\Rightarrow$ (b): This is obvious.

(b) $\Rightarrow$ (c): Suppose that f is not bounded. Then there exists a bounded subset B of E such that the set $\{f(x);\ x \in B\}$ is unbounded in F. But then there exist sequences $\{y_n\} \subset B$ and $\{\lambda_n\} \subset \mathbb{R}$, $\lambda_n \to 0$ such that $\lambda_n f(y_n) \not\to 0$. Putting $x_n = \lambda_n y_n \to 0$, we arrive at a contradiction.

(c) $\Rightarrow$ (a): This follows from Theorem 13.

Kōmura [30] has shown that there exist reflexive spaces which are not complete; this shows that the strong dual of a barrelled space need not be complete. However, for bornological spaces, it turns out to be true as shown below.

Proposition 14. If (E,u) is a bornological space, $(E',\beta(E',E))$ is complete.

Proof: Every $\beta(E',E)$-Cauchy net in E' converges pointwise to a linear functional f on E; f is bounded, since the convergence on each bounded subset of E is uniform. Hence $f \in E'$ by Corollary 16. This proves that $(E',\beta(E',E))$ is complete.

Corollary 17. Let (E,u) be a metrizable l.c. space. The following statements are equivalent:

(a) $(E',\beta(E',E))$ is bornological.

(b) $(E',\beta(E',E))$ is quasibarrelled.

(c) $(E',\beta(E',E))$ is barrelled.

Proof: (a) $\Rightarrow$ (b): This is always true, in view of Proposition 11.

(b) $\Rightarrow$ (c): This is obtained by combining Corollary 6 and Proposition 14.

(c) $\Rightarrow$ (a): Let $\{B_n\}$ be a fundamental sequence of circled, convex, $\sigma(E',E)$-compact and equicontinuous sets. Since each B_n is strongly bounded, for any circled, convex and bornivorous set A in E' there exists $\lambda_n > 0$ such that $2\lambda_n B_n \subset A$

Let A_n be the convex hull of $\bigcup_{k=1}^{n} \lambda_k B_k$. Then each A_n is circled, convex and $\sigma(E',E)$-compact, and $A_0 = \bigcup_{n=1}^{\infty} A_n$ is a circled, convex and absorbing set such that $2A_0 \subset A$. We claim that $cl_\beta(A_0) \subset 2A_0$. Let $x' \notin 2A_0$. Since A_n is $\beta(E',E)$-closed, for each $n \geq 1$ there exists a circled convex β-neighbourhood V_n in E' such that $(x' + V_n) \cap A_n = \emptyset$. Let $W_n = V_n + A_0$. Then, by (Chapter 1, Proposition 27), $W = \bigcap_{n=1}^{\infty} W_n$ is a neighbourhood of 0 in $(E', \beta(E',E))$. On the other hand, $x' \notin 2A_0$ implies that $(x' + W_n) \cap A_n = \emptyset$ for all n, whence $(x' + W) \cap A_0 = \emptyset$. We now conclude that $x' \notin cl_\beta(A_0)$. Thus we have proved that $cl_\beta(A_0) \subset 2A_0$. But $cl_\beta(A_0)$ is a neighbourhood of 0 in $(E', \beta(E',E))$ and so is A_0. This completes the proof.

There exist bornological spaces whose duals are not bornological, e.g. see Köthe [31], §31 (7).

Let (E,u) be an l.c. space and $\mathcal{B}$ the family of all bounded subsets of (E,u). Let u_0 be the finest l.c. topology on E whose family of bounded sets is identical with $\mathcal{B}$.

Proposition 15. (E,u_0) is a bornological space.

Proof: The circled convex neighbourhoods of 0 for u_0 absorb all the members of $\mathcal{B}$. On the other hand, the family of all circled, convex and bornivorous sets form a base of neighbourhoods of 0 for an l.c. topology on E, and this must be u_0. We conclude that (E,u_0) is bornological.

Corollary 18. An l.c. space (E,u) is bornological iff $u = u_0$.

Definition 4. (E,u_0) is called the bornological space associated with (E,u).

Proposition 16. Every (sequentially complete) bornological space (E,u) is the inductive limit of a family of (complete) normed spaces.

Proof: Let $\mathcal{B}$ denote the family of all closed, circled, convex and bounded subsets of (E,u), ordered by inclusion. For a fixed $B \in \mathcal{B}$, consider the subspace E_B of E defined by

$$E_B = \bigcup_{n=1}^{\infty} n\, B.$$

Let p_B be the gauge of B in E_B. Since B is bounded and (E,u) is Hausdorff, it follows that p_B is a norm on E_B and u induces a coarser topology on E_B than the norm topology. Evidently, the inductive limit topology on E with respect to $\{[(E_B,p_B): g_B];\ B \in \mathcal{B}\}$, where g_B is the identity map of E_B into E, is the bornological topology u_0 associated with u and hence by Corollary 18, $u = u_0$. This completes the proof of one part. The proof of the other part follows immediately since (E,u) being sequentially complete implies (E_B,p_B) is complete.

Proposition 17. The inductive limit of a family of bornological spaces is again bornological.

Proof: Let E be a vector space, $\{(E_\alpha, u_\alpha): \alpha \in A\}$ a family of bornological spaces and f_α a linear map of E_α into E for each α. Let u be the inductive limit topology on E with respect to $\{[(E_\alpha, u_\alpha): f_\alpha]: \alpha \in A\}$. We show that (E,u) is bornological. Let V be a circled, convex and bornivorous set in E. Then $f_\alpha^{-1}(V)$ is a circled, convex and bornivorous set in E_α and hence a neighbourhood of 0 in E_α. Hence by the definition of inductive limits, V is a neighbourhood of 0.

Corollary 19. Every quotient of a bornological space is bornological.

Corollary 20. The locally convex direct sum of a family of bornological spaces is bornological.

A closed subspace of a bornological space need not be bornological ([31], §28 (4)). However, every finite codimensional subspace of a bornological space is bornological ([31], §28 (4)). The proof of the following proposition is postponed until Chapter 3, Proposition 11.

Proposition 18. Every countable product of bornological spaces is bornological.

A cardinal d_0 is called strongly inaccessible if

(i) $\aleph_0 < d_0$,

(ii) $\Sigma\{d_\alpha : \alpha \in A\} < d_0$,

whenever card(A) $< d_0$ and $d_\alpha < d_0$ for all $\alpha \in A$, and

(iii) $d < d_0$

implies that $2^d < d_0$. It is unknown if such cardinals exist. The Mackey-Ulam Theorem ([31], §28(8)) asserts that every product of d bornological spaces is bornological if d is smaller than the smallest strongly inaccessible cardinal. For more information, see Köthe ([31], 28(8)). Kōmura [29] has shown that an arbitrary l.c. space with dimension $<d_0$ is a closed subspace of some bornological space.

§3. Distinguished spaces.

Definition 5. An l.c. space E is said to be a distinguished space if each $\sigma(E'',E')$-bounded subset of its strong bidual E" is contained in the $\sigma(E'',E')$-closure of some bounded subset of E.

This is the same as saying that for each $\sigma(E'',E')$-bounded subset B of E", there is a bounded subset A of E such that $B \subset A^{oo}$, where A^{oo} is the polar of A^o with respect to the dual pair $\langle E',E''\rangle$.

Every normed space is clearly a distinguished space. Also, every semi-reflexive space is distinguished.

Theorem 14. An l.c. space E is distinguished iff $(E',\beta(E',E))$ is barrelled.

Proof: Assume that E is distinguished. Let B be a $\sigma(E'',E')$-bounded subset of E". Then there exists a bounded subset A of

E such that $B \subset A^{oo}$. But A^o is a neighbourhood of 0 in $(E', \beta(E',E))$ so that A^{oo} is equicontinuous. This shows that B is equicontinuous and hence $(E', \beta(E',E))$ is barrelled. Conversely, assume that $(E', \beta(E',E))$ is barrelled. Let B be a $\sigma(E'',E')$-bounded subset of E'. Then B is equicontinuous so that $B \subset U^o$ for some neighbourhood U of 0 in $(E', \beta(E',E))$. It now follows that $B \subset U^o \subset A^{oo}$ for some bounded set A in E. We conclude that E is distinguished.

A distinguished space need not be quasibarrelled (see Chapter 5, §5, Example 9). There exist complete metrizable l.c. spaces (and hence barrelled and quasibarrelled spaces) which are not distinguished. The following result which follows from Theorem 14 and Corollary 17 tells us when a metrizable space is distinguished.

Corollary 21. A metrizable l.c. space E is distinguished iff $(E', \beta(E',E)$ is bornological.

Proposition 19. If E is a distinguished space, then E" is a $\sigma(E,E')$-quasicompletion of E.

Proof: Let B be any bounded subset of E". Then there exists a bounded subset A of E such that B is contained in A^{oo} which is $\sigma(E'',E')$-compact, and so $\sigma(E'',E')$-complete.

For the proof of the following theorem we refer the reader to ([11], page 291).

Theorem 15. Let $\{E_n\}$ be a sequence of subspaces of a vector space E such that $E_n \subset E_{n+1}$ for all $n \geq 1$ and $E = \bigcup_{n=1}^{\infty} E_n$. Let each E_n be equipped with an l.c. topology u_n such that u_{n+1} induces u_n on E_n. If each (E_n, u_n) is metrizable and distinguished, then E, equipped with the inductive limit topology, is distinguished and $(E', \beta(E',E))$ is bornological.

§4. (DF)-spaces.

Motivated by two interesting properties of strong duals of metrizable l.c. spaces (Chapter 1, Propositions 26 and 27), Grothendieck [10] introduced the concept of (DF)-space . All the results of this section, unless otherwise stated, are due to Grothendieck.

Definition 6. An l.c. space E is called a (DF)-space if (i) it has a fundamental sequence of bounded sets and (ii) each $\beta(E',E)$-bounded subset of E', which is the countable union of equicontinuous subsets of E' is itself equicontinuous.

As will be seen in Chapter 5, the condition (ii) in the above definition is equivalent to (ii)' each bornivorous barrel which is the countable intersection of closed, circled, convex neighbourhoods of 0 (which is called bornivorous N-barrel in Chapter 5) is itself a neighbourhood of 0.

It is easy to check that every normed space is a (DF)-space and the strong dual of a metrizable l.c. space is a complete (DF)-space. A (DF)-space need not be quasibarrelled ([31], §31 (7)).

Theorem 16. Let E be a (DF)-space. A circled convex subset U of E is a neighbourhood of 0 iff for each circled, convex bounded subset A of E, $A \cap U$ is a neighbourhood of 0 in A.

Proof: The condition is evidently necessary. To prove the sufficiency, let $\{A_n\}$ be a fundamental sequence of closed, circled convex bounded subsets of E. We construct a sequence $\{\lambda_n\}$ of positive numbers and a sequence $\{V_n\}$ of closed, circled convex neighbourhoods of 0 such that for all i, j

(i) $\lambda_i A_i \subset (1/3)U$,

(ii) $\lambda_i A_i \subset V_j$ and

(iii) $V_i \cap A_j \subset U$.

Assume that (i)-(iii) are true for all $i,j \leq n$. We choose $\lambda_{n+1} > 0$ such that $\lambda_{n+1}A_{n+1} \subset (1/3)U$; this is possible, since, by hypothesis, there exists a neighbourhood V of 0 such that $V \cap A_{n+1} \subset U$ and λ_{n+1} can be so chosen that $\lambda_{n+1}A_{n+1} \subset (1/3)V$ and $\lambda_{n+1}A_{n+1} \subset (1/3)A_{n+1}$. Thus (i) is satisfied for n+1. Furthermore, λ_{n+1} can be chosen so small enough that (ii) holds for $i = n+1$ and $j \leq n$. The construction will be complete if we find a closed, circled, convex neighbourhood V_{n+1} of 0 such that

$$\lambda_i A_i \subset V_{n+1} \quad \text{for } i \leq n+1 \text{ and } V_{n+1} \cap A_{n+1} \subset U.$$

Now, let $A^{(n+1)} = \Gamma_{i=1}^{n+1} \lambda_i A_i$ (circled convex hull of $\lambda_i A_i$, $i = 1,2,\ldots n+1$) and $V_{n+1} = \overline{A^{(n+1)} + V}$, where V is a circled convex neighbourhood of 0 to be determined. Since $A^{(n+1)} \subset V_{n+1} \subset 2(A^{(n+1)} + V) = 2A^{(n+1)} + 2V$, it is sufficient to show that $(2A^{(n+1)} + 2V) \cap A_{n+1} \subset U$.

If we write $B = A_{n+1} \cap (E \setminus U)$, we must show that $(2A^{(n+1)} + 2V) \cap B = \emptyset$ or $2V \cap (B + 2A^{(n+1)}) = \emptyset$. This means that we must show that 0 can not be a closure point of $C = B + 2A^{(n+1)}$. To prove this, we first observe that, since $A^{(n+1)} \subset 1/3U$, $1/3U + 2A^{(n+1)} \subset 1/3U + 2/3U \subset U$. Since $U \cap B$ is empty, it follows that $(1/3U + 2A^{(n+1)}) \cap B = \emptyset$ and so $(3\,C) \cap U$ is empty. The set $3\,C$ is bounded and so $3\,C \subset A_j$ for some j; also $U \cap A_j$ is a neighbourhood of 0 in A_j. Since $(3\,C) \cap U = \emptyset$, 0 is not a closure point of $3\,C$ and so not a closure point of C either. Now we conclude that $W = \bigcap_{i=1}^{\infty} V_i$ is a neighbourhood of 0 in view of (ii) and the definition of (DF)-space and that $W \subset U$ in view of (iii) and the fact that $UA_n = E$.

A simple consequence of Theorem 16 is the following:

<u>Corollary 22</u>. A linear map f of a (DF)-space E into an l.c. space F is continuous iff its restriction to each bounded subset of E is continuous.

<u>Corollary 23</u>. Let E be a (DF)-space. Then $(E', \beta(E',E))$ is a Fréchet (or complete metrizable l.c.) space.

Proof: Clearly $(E', \beta(E',E))$ is metrizable; completeness can be established using Corollary 22.

Proposition 20. Let (E,u) be a (DF)-space, M a separable subset of E and u_0 the topology on E of uniform convergence on the strongly bounded subsets of E'. Then u_0 induces on M the topology of E.

Proof: Since u_0 is finer than u it is sufficient to show that for each closed, circled, convex, u_0 neighbourhood U of 0, there exists an open neighbourhood V of 0 in E such that $V \cap M \subset U$ or $(V \cap (E\setminus U)) \cap M = \emptyset$. Since V is open, it is sufficient to show that $x_i \notin V \cap (E\setminus U)$ for all i, where $\{x_i\}$ is a dense sequence in M. In other words, we must show that for each sequence $\{x_{i_k}\}$, $x_{i_k} \neq 0$ which we again write $\{x_i\}$ in $E\setminus U$, there exists a neighbourhood V of 0 in E such that $x_i \notin V$ for all i. For this, we construct a sequence of closed, circled and convex neighbourhoods V_n of 0 and a sequence $\{\lambda_n\}$ of positive numbers such that

(a) $\lambda_i A_i \subset V_j$,

(b) $\lambda_i A_i \subset U$, and

(c) $x_i \in E\setminus V_i$ for all i and j,

where $\{A_n\}$ is a fundamental sequence of bounded sets in E. Assume the validity of (a)-(c) for $i,j \leq n$. Clearly it is possible to choose λ_{n+1} in such a way that (b) remains true for n+1 and (a) for $i = n+1$ and $j \leq n$. Let $A^{(n+1)} = \Gamma_{i=1}^{n+1} \lambda_i A_i$. Since $A^{(n+1)} \subset U$ and $x_{n+1} \notin U$, it is possible to choose a closed circled convex neighbourhood V_{n+1} of 0, $A^{(n+1)} \subset V_{n+1}$, such that $x_{n+1} \notin V_{n+1}$. Let $V = \bigcap_{i=1}^{\infty} V_i$. Then (a) implies that V is a bornivorous N-barrel and hence a neighbourhood of 0 in E. By (c), we conclude that $x_i \notin V$.

Corollary 24. The u_0-convergent sequences are the same as u-convergent sequences and u_0 coincides with u on the metrizable subsets of E.

Corollary 25. A separable (DF)-space is quasibarrelled.

Remark 2. Corollary 25 also follows from (Chapter 5, Remark 1 and Corollary 6).

Corollary 26. Let (E,u) be a (DF)-space. If the bounded subsets of (E,u) are metrizable, then (E,u) is quasibarrelled.

Proof: In view of Theorem 16, it is sufficient to show that u_0 coincides with u on the bounded subsets of E. But this follows from Corollary 24.

The proofs of the following interesting properties of (DF)-spaces are omitted; the interested reader is referred to Köthe [31] and Grothendieck [10] for their proofs:

(1) A (DF)-space is complete iff it is quasi-complete.

(2) The completion of a (DF)-space is again a (DF)-space.

Remark 3. (2) follows easily from (Chapter 5, Proposition 3).

(3) The quotient of a (DF)-space by a closed subspace is

again a (DF)-space.

(4) The l.c. direct sum and the inductive limit of a sequence of (DF)-spaces are (DF)-spaces.

A closed subspace of a (DF)-space need not be a (DF)-space ([31], §31 (5). See also Chapter 5, Example 5). However, we show below that a finite codimensional subspace of a (DF)-space is a (DF)-space. This result is due to Valdivia [53].

Lemma 7. Let (E,u) be an l.c. space and M a subspace of E of codimension one. If V is a bornivorous barrel in M such that its closure $\overline{V}$ in E is absorbing in E, then $\overline{V}$ is bornivorous in E.

Proof: Let $\mathfrak{B}$ be the family of all closed, circled, convex and bounded sets in E. If $B \in \mathfrak{B}$, we write E_B to denote the linear hull of B with the norm associated with B. Let (E,v) be the inductive limit topology of $\{E_B\colon B \in \mathfrak{B}\}$. Clearly $u \subset v$. Furthermore, (E,v) is bornological because each E_B being a normed space is bornological and the inductive limit of bornological spaces is bornological. Hence M with the topology induced by v, which we write (M,v), is bornological. Hence V is a neighbourhood of 0 in (M,v). Let $\overline{V}$ and $\overline{V}^*$ be the closures of V in (E,u) and (E,v) respectively. If $\overline{V}^*$ is absorbing in E, then it is a neighbourhood of 0 in (E,v) and hence bornivorous in (E,u), since u-bounded sets are the same as v-bounded sets. But then $\overline{V}$ is bornivorous in (E,u) because $\overline{V} \supset \overline{V}^*$. If $\overline{V}^*$ is not absorbing in (E,u), then $V = \overline{V}^*$. Let $x \in \overline{V} \setminus V$ and let C be the circled hull of $\{x\}$. Then V+C is a neighbourhood of 0 in (E,v). Then it will be bornivorous in (E,u). On the other hand, $\overline{V} \supset V$ and $\overline{V} \supset C$ and so $2\overline{V}$ (or $\overline{V}$) is bornivorous in (E,u). This completes the proof.

<u>Proposition 21</u>. Let E be a (DF)-space and M a finite codimensional subspace of E. Then M is a (DF)-space.

<u>Proof</u>: It suffices to prove the result when M is of codimension one. First we assume that M is closed in E. Let $x \in E\backslash M$ and let L be the linear hull of $\{x\}$. Then M is isomorphic to E/L and hence a (DF)-space. Next we assume that M is dense in E. Let $V = \bigcap_{n=1}^{\infty} V_n$ be a bornivorous N-barrel in M. Let $\{B_n;\ n \geq 1\}$ be a fundamental sequence of closed, circled, convex and bounded sets in E. Since $(E',\beta(E',E))$ is complete, there exists an integer $n_0 > 0$ such that $M \cap B_{n_0}$ is not closed in E. Hence there exists an $x \in \overline{M \cap B_{n_0}} \backslash M \cap B_{n_0}$ and so $x \notin M$. Clearly $\lambda(M \cap B_{n_0}) \subset V$ for some $\lambda > 0$, and $\lambda x \in \overline{V}$. But $\lambda x \notin M$ and M is a hyperplane in E. Thus $\overline{V}$ is absorbing in E, and hence by Lemma 7, $\overline{V}$ is bornivorous in E. But then $\bigcap_{n=1}^{\infty} \overline{V}_n$ is bornivorous in E and so a neighbourhood of 0 in E. Since $(\bigcap_{n=1}^{\infty} \overline{V}_n) \cap M = \bigcap_{n=1}^{\infty} (\overline{V}_n \cap M) = \bigcap_{n=1}^{\infty} V_n = V$, it follows that M is a (DF)-space. The general case is now obvious. This completes the proof.

<u>Remark 4</u>. For the case of a countable codimensional subspace

of a (DF)-space, see (Chapter 5, Corollary 18) and (Chapter 6, Corollary 11).

Let E and F be l.c. spaces and $\mathcal{B}_1$ and $\mathcal{B}_2$ the families of all bounded subsets of E and F respectively. The topology on $\mathcal{L}(E,F)$ of bi-bounded convergence, i.e. $\mathcal{B}_1 \times \mathcal{B}_2$-topology, is coarser than $\beta(\mathcal{L}(E,F), E \tilde{\otimes}_p F)$. The following theorem, due to Grothendieck, shows when the topologies coincide. The proof is omitted, since it is similar to that of (Chapter 8, Theorem 6).

Theorem 17. Let E and F be (DF)-spaces. Then the topology on $\mathcal{L}(E,F)$ of bi-bounded convergence coincides with $\beta(\mathcal{L}(E,F), E \tilde{\otimes}_p F)$.

CHAPTER III

ULTRABARRELLED, ULTRABORNOLOGICAL AND QUASIULTRABARRELLED SPACES

Robertson [44] introduced the concept of ultrabarrelled space to replace that of barrelled space in non-locally convex situations. Iyahen [19] characterized ultrabarrelled spaces in terms of what he calls ultrabarrels and obtained some interesting results for ultrabarrelled spaces. He also introduced the concepts of ultrabornological and quasiultrabarrelled spaces to replace those of bornological and quasibarrelled spaces in non-locally convex situations. (We remark that the concept of ultrabornological space considered by Bourbaki [4] is different from the one considered here.) In this chapter we deal with these three concepts. Almost all the results of this chapter are due to Robertson [44], Iyahen [19] and Adasch [1]. See also [54].

§1. Ultrabarrelled and quasiultrabarrelled spaces.

Definition 1. Let (E,u) be a t.v. space. A closed circled subset B of E is called an ultrabarrel (a bornivorous ultrabarrel) if there exists a sequence $\{B_n;\ n \geq 1\}$ of closed,circled,absorbing (closed,circled,bornivorous) subsets of E such that $B_1 + B_1 \subseteq B$ and $B_{n+1} + B_{n+1} \subseteq B_n$ for all $n \geq 1$.

The sequence $\{B_n\}$ is called a defining sequence for B.

Remark 1. If the closedness of B and B_n's is dropped in the above definition, then B is called suprabarrel (bornivorous suprabarrel).

If B is a barrel (bornivorous barrel) in a t.v. space E, then B is an ultrabarrel (bornivorous ultrabarrel) with $\{2^{-n}B\}$ as a defining sequence. However, an ultrabarrel (a bornivorous ultrabarrel) need not be convex and need not have a defining sequence of convex sets. (For an example, consider a complete non-locally convex locally bounded t.v. space and B a closed, circled, bounded neighbourhood of 0 with $\{\lambda_n B\}$, $\lambda_n > 0$ as a defining sequence.)

The following proposition can easily be verified.

Proposition 1. (a) Any closed, circled neighbourhood of 0 in a t.v. space is a bornivorous ultrabarrel. (b) Let E and F be t.v. spaces and f a linear map of E into F. Then

(i) $\overline{f^{-1}(B)}$ is an ultrabarrel (a bornivorous ultrabarrel) in E whenever B is an ultrabarrel in F (B is a bornivorous ultrabarrel in F and f is bounded). If f is continuous, then $f^{-1}(B)$ is an ultrabarrel (a bornivorous ultrabarrel) in E whenever B is an ultrabarrel (a bornivorous ultrabarrel) in F;

(ii) If f is onto, then $\overline{f(A)}$ is an ultrabarrel in F, whenever A is an ultrabarrel in E.

Definition 2. A t.v. space (E,u) is called ultrabarrelled if any linear topology on E with a base of u-closed neighbourhoods of 0 is necessarily coarser than u.

Definition 3. A t.v. space (E,u) is called quasiultrabarrelled iff each bornicorous ultrabarrel in E is a neighbourhood of 0.

Proposition 2. Every Baire t.v. space is ultrabarrelled.

Proof: Let (E,u) be a Baire t.v. space and v any linear topology on E which has a base of u-closed neighbourhoods of 0. Let V be any v-neighbourhood of 0. We show that V is an u-neighbourhood of 0 and this will complete the proof. There exists a v-neighbourhood W of 0 such that W is u-closed circled and $W+W \subset V$. Since $E = \bigcup_{n=1}^{\infty} n W$ is a Baire space, there is an n such that the interior of nW (and so of W) in (E,u) is not empty. Let x_0 be an u-interior point of W. Then $0 = x_0 - x_0 \in W+W \subset V$ and so V is an u-neighbourhood of 0.

Corollary 1. Every complete metrizable t.v. space is ultrabarrelled.

Theorem 1. A t.v. space (E,u) is ultrabarrelled iff each ultrabarrel in E is a neighbourhood of 0.

Proof: This follows from the definition, since (E,u) is ultrabarrelled iff each linear topology with a base of neighbourhoods of 0 consisting of u-ultrabarrels is coarser than u.

Remark 2. Clearly an ultrabarrelled space is quasiultrabarrelled.

Corollary 2. Let (E,u) and (F,v) be t.v. spaces and f a linear map of E into F.

(a) If (E,u) is ultrabarrelled, then f is almost continuous.

(b) If f is onto and (F,v) is ultrabarrelled, then f is almost open.

Proof: This follows from Theorem 1 and Proposition 1(b).

Remark 3. If there is a linear continuous almost open map of an l.c. space (E,u) onto a t.v. space F, then F is an l.c. space. Hence, by Corollary 2, it follows that a non-locally convex ultrabarrelled topology on E cannot be coarser than u. Since the finest vector topology on a countable dimensional vector space is locally convex, any countable dimensional ultrabarrelled space is necessarily locally convex.

Corollary 3. Let (E,u) be an ultrabarrelled (quasiultrabarrelled) space and (F,v) any t.v. space. If f is a linear continuous almost open map of (E,u) into (F,v), then (F,v) is ultrabarrelled (quasiultrabarrelled).

Proof: This again follows from Theorem 1, Definition 3 and Proposition 1.

An immediate consequence of Corollary 3 is the following:

Corollary 4. A quotient by a subspace of an ultrabarrelled (quasiultrabarrelled) space is ultrabarrelled (quasiultrabarrelled).

Corollary 5. A t.v. space (E,u) is ultrabarrelled iff the identity map of (E,τ) onto (E,u) is almost open, where τ is the finest vector topology on E.

Proof: Since (E,τ) is clearly ultrabarrelled, the result follows from Corollaries 2 and 3.

Proposition 3. Let (E,u) be an ultrabarrelled space and F a subspace of $(\tilde{E},\tilde{u})$ such that $E \subset F \subset \tilde{E}$ where $\tilde{E}$ is the completion of E. Then F is ultrabarrelled.

Proof: Let v be any vector topology on F with a base of neighbourhoods $\{U\}$ of 0 such that each U is $\tilde{u}$-closed in F. Then v induces a topology with a base of $\tilde{u}$-closed neighbourhoods of 0 in E. For each U, clearly $U \cap E$ is an $\tilde{u}$-neighbourhood of 0 in E. Hence $\widetilde{U \cap E}$ and so $\tilde{U}$ is an $\tilde{u}$-neighbourhood of 0 in $\tilde{E}$. But $U = \tilde{U} \cap F$, since U is $\tilde{u}$-closed in F and hence U is an $\tilde{u}$-neighbourhood of 0 in F. Thus v is coarser than the topology $\tilde{u}$ on F which proves that F is ultrabarrelled.

Corollary 6. The completion of an ultrabarrelled space is ultrabarrelled.

Lemma 1. An ultrabarrel B in a t.v. space (E,u) absorbs every circled, semiconvex, sequentially complete and bounded set.

Proof: Let M be a circled, semi-convex, sequentially complete and bounded set in (E,u). We assume, without loss of generality, that M spans E. Let v be the locally bounded vector topology on E with M as a bounded neighbourhood of 0. Then v is finer than u and so B is an ultrabarrel in (E,v). Furthermore, since M is sequentially complete, it follows that (E,v) is complete. Hence, (E,v) is ultrabarrelled. But then B is a neighbourhood of 0 in (E,v) and so it absorbs M.

Remark 4. As in the locally convex case, we can prove that an ultrabarrel in a t.v. space absorbs each circled, convex and compact set.

Proposition 4. A sequentially complete almost convex quasi-ultrabarrelled space (E,u) is ultrabarrelled.

Proof: Let V be an ultrabarrel in (E,u). If A is a bounded set in (E,u), then $A \subset B$ for some closed, circled, semi-convex and bounded set B. Clearly B is sequentially complete. Hence, by Lemma 1, V absorbs B and so absorbs A as well. Thus V is a bornivorous ultrabarrel in (E,u) and hence a neighbourhood of 0. This proves that (E,u) is ultrabarrelled.

Theorem 2. Let (E,u) be a t.v. space. The following statements are equivalent:

(a) (E,u) is ultrabarrelled (quasiultrabarrelled).

(b) Each linear (and bounded) map of E into a complete metrizable t.v. space F, with closed graph, is continuous.

(c) Each lower semi-continuous F-seminorm p on E is continuous (when $\sup\{p(sx);\ x \in B\} \to 0$ as $s \to 0$ for all $B \in G$, where G is the family of all bounded sets in E).

Proof: We prove the theorem for ultrabarrelled spaces only. The other part is similar.

(a) $\Rightarrow$ (b): This follows from Corollary 2(a) and (Chapter 1, Theorem 7(a)).

(b) $\Rightarrow$ (c): Let p be a lower semi-continuous $\mathfrak{F}$-seminorm on E and $(F,\tilde{p})$ the completion of the $\mathfrak{F}$-normed vector space $(E/p^{-1}(0),p)$. We show that the canonical map ϕ: $E \to F$ has closed graph. Let $\{x_\lambda\}$ be a net in E converging to x with respect to the topology u and $\phi(x_\lambda)$ converging to y (in F) with respect to the F-norm topology $\tilde{p}$. Then for any $\varepsilon > 0$ there exists $x' \in E$ such that $\tilde{p}(y - \phi(x')) \leq \varepsilon/2$. Then

$$\begin{aligned}
\tilde{p}(y - \phi(x)) &\leq \tilde{p}(y - \phi(x')) + \tilde{p}(\phi(x') - \phi(x)) \\
&\leq \tilde{p}(y - \phi(x')) + p(x' - x) \\
&\leq \tilde{p}(y - \phi(x')) + \lim\inf p(x' - x_\lambda) \\
&\leq \tilde{p}(y - \phi(x') + \lim \tilde{p}(\phi(x') - \phi(x_\lambda)) \\
&\leq 2\tilde{p}(y - \phi(x')) \leq \varepsilon.
\end{aligned}$$

This shows that $y = \phi(x)$. It now follows that p is continuous.

(c) $\Rightarrow$ (a): Let B_0 be an ultrabarrel in (E,u) with $\{B_n\}$ as a defining sequence. Let $\alpha = q/_2N$ be a dyadic rational, $0 \leq \alpha < 1$. Then $\alpha = \sum_1^N t_k 2^{-k}$ with $t_k = 0$ or 1 for all k. Define $U_\alpha = \sum_1^N t_k B^k$. Let $\bar{U}_\alpha$ be the u-closure of U_α. Define $p(x) = \inf\{\alpha > 0 : x \in \bar{U}\}$. Then p is a lower semi-continuous $\mathfrak{F}$-seminorm and so p is continuous. It now follows from $B_n = \{x: p(x) \leq 2^{-n}$ that B_0 is a neighbourhood of 0.

<u>Corollary 7</u>. Any *-inductive limit of ultrabarrelled (quasi-ultrabarrelled) spaces is ultrabarrelled (quasiultrabarrelled).

<u>Proof</u>: Let (E,u) be the *-inductive limit of $\{[(E_\alpha, u_\alpha): f_\alpha]; \alpha \in A\}$ where each (E_α, u_α) is ultrabarrelled. If g is a linear map of E into a complete metrizable t.v. space F, with closed graph, then it follows that $g o f_\alpha$ is a linear map of E_α into F, with closed graph, for each α. Hence, by Theorem 2, $g o f_\alpha$ is continuous for all α which implies that g is continuous. It now follows from Theorem 2 that E is ultrabarrelled.

As a particular case we have

<u>Corollary 8</u>. The *-direct sum of any family of ultrabarrelled (quasiultrabarrelled) spaces is ultrabarrelled (quasiultra-barrelled).

<u>Corollary 9</u>. Any countable inductive limit of l.c. ultrabarrelled spaces is ultrabarrelled. In particular, any countable inductive limit of Fréchet spaces is ultrabarrelled.

The following result is due to Adasch [1].

<u>Proposition 4'</u>. Any product of ultrabarrelled (quasiultrabarrelled) spaces is ultrabarrelled (quasiultrabarrelled).

<u>Proof</u>: Let $\{(E_\alpha, u_\alpha): \alpha \in A\}$ be a family of ultrabarrelled spaces and $(E,u) = \prod_{\alpha \in A} (E_\alpha, u_\alpha)$. We need to prove the result when A is infinite. Let V be an ultrabarrel in (E,u) with $\{V_n\}$ as a defining sequence. First we show that

(a) there exists a finite subset $\mathcal{B} \subset A$ such that $V_1 \supset \prod_{A\setminus} E_\alpha$. Since V_1 is closed and the closure of the direct sum $\bigoplus_{A\setminus\mathcal{B}} E_\alpha$ is $\prod_{A\setminus\mathcal{B}} E_\alpha$, it is sufficient to show that

(b) there exists a finite subset $\mathcal{B} \subset A$ such that $V_1 \supset \bigoplus_{A\setminus\mathcal{B}} E_\alpha$. Assume that (b) is false. Then there exists an $x^{(1)} \in \bigoplus_A E_\alpha$ such that $x^{(1)} \notin V_1$. Let $Tr(x^{(1)}) = \{\alpha \in A: x_\alpha^{(1)} \neq 0\}$. Since $T(x^{(1)})$ is a finite subset of A, there exists an $x^{(2)} \in \bigoplus_{A\setminus T(x^{(1)})} E_\alpha$ with $x^{(2)} \notin 2V_1$. We therefore obtain, by induction, a sequence $x^{(n)} \in \bigoplus_{A\setminus\bigcup_{i=1}^{n-1} T(x^{(i)})} E_\alpha$ with $x^{(n)} \notin nV_1$, $n \geq 1$.

(b_1) In each $\prod_{T(x^{(i)})} E_\alpha$, consider the circled, convex and compact subset $K_i = \{(\lambda_\alpha x_\alpha^{(i)})_{\alpha \in Tr(x^{(i)})}: |\lambda_\alpha| \leq 1\}$. Since $K_i \cap K_j = 0$ for $i \neq j$, we can consider $\prod_{i=1}^{\infty} K_i = K$ in $\prod_A E_\alpha$. K is then circled, convex and compact in $\prod_A E_\alpha$. But, since $x^{(n)} \in K$ and $x^{(n)} \notin nV_1$, the ultrabarrel V_1 would not absorb K which is a contradiction by Remark 4. This proves (b) and hence (a). We now obtain the result from (a) as follows: Clearly $V_1 \cap \prod_{\mathcal{B}} E_\alpha$ is an ultrabarrel in $\prod_{\mathcal{B}} E_\alpha$ with $\{V_i \cap \prod_{\mathcal{B}} E_\alpha: i \geq 2\}$ as a defining sequence. By Corollary 8, $V_1 \cap \prod_{\mathcal{B}} E_\alpha$ is a neighbourhood of 0 in $\prod_{\mathcal{B}} E_\alpha$. Hence, $V \supset V_1 + V_1 \supset V_1 \cap \prod_{\mathcal{B}} E_\alpha + \prod_{A\setminus\mathcal{B}} E_\alpha$ is also a neighbourhood of 0 in $\prod_A E_\alpha$. This completes the proof.

Similarly we can show that any product of quasiultrabarrelled spaces is quasiultrabarrelled.

Remark 5. The proof of Proposition 4 can be adapted to show that a product of barrelled (quasibarrelled) spaces is barrelled (quasibarrelled). The proof can be simplified by taking $\frac{1}{2}V$, V barrel, instead of V_1. To prove (b), it is sufficient to show that $\frac{1}{2}V \supset E_\alpha$ for $\alpha \in A\setminus B$ which simplifies part (b_1).

The following example shows that a closed subspace of an ultrabarrelled space need not be ultrabarrelled.

Example 1. Every Hausdorff t.v. space is a subspace of a product of metrizable t.v. spaces ([28], Theorem 1), and it is an easy consequence of this result that any complete Hausdorff t.v. space is a closed subspace of a product of complete metrizable t.v. spaces which is non-meagre ([4], §4, Exercise 7) and hence ultrabarrelled. Since an uncountable dimensional vector space under its finest l.c. topology is not ultrabarrelled but is complete and Hausdorff, it follows that a closed subspace of an ultrabarrelled space need not be ultrabarrelled.

Proposition 5. A finite codimensional subspace of an ultrabarrelled space is ultrabarrelled.

Proof: Let E be an ultrabarrelled space and M a finite codimensional subspace of E. Let f be a linear map of M into a complete metrizable t.v. space F, with graph G closed in $M \times F$. Let G_1 be the closure of G in $E \times F$. Let $(0,y) \in G_1$. Then

$(0,y) \in G_1 \cap (M \times F) = G$, because G is closed in $M \times F$ and hence $y = 0$. Since G_1 is a vector subspace of $E \times F$, it follows that G_1 is the graph of a linear extension f_1 of f from a vector subspace of E, containing M, into F. Let f_2 be a linear extension of f_1 to all of E into F. The graph G_2 of f_2 can be expressed in the form $G_2 = G_1 + F_1$, where F_1 is a finite-dimensional subspace of $E \times F$. Thus G_2 is closed in $E \times F$ and so f_2 is continuous by Theorem 2. Hence f is continuous and again by Theorem 2 it follows that M is ultrabarrelled.

Proposition 6. Let (E,u) be an ultrabarrelled space and (F,v) any complete metrizable t.v. space. If f is a continuous linear map of F onto E, then it is open.

Proof: If V is a closed, circled neighbourhood of 0 in F, then $\overline{f(V)}$ is an ultrabarrel in E by Proposition 1(b), part (ii) and hence a neighbourhood of 0 in E. This implies that f is almost open. It now follows from (Chapter I, Theorem 7(b)), that f is open.

Definition 3. Let E and F be t.v. spaces. (a) Let f be a linear map of E into F. We say that the filter condition holds if for any Cauchy filter base $\mathfrak{F}$ on E such that $f(\mathfrak{F})$ is convergent to a point of f(E), it follows that $\mathfrak{F}$ is convergent to a point of E. (b) Let g be a linear map of F into E. We say that the inverse filter condition holds if for any convergent filter base $\mathfrak{F}$ on F such that $g(\mathfrak{F})$ is Cauchy, it follows that

$g(\mathfrak{F})$ is convergent to a point in $g(F)$.

Lemma 2. If u and v are compatible topologies with the duality of the same dual pair $\langle E,E'\rangle$ and if $(\tilde{E}_u,E')$ is also a dual pair, then the filter condition holds.

Proof: Let $\mathfrak{F}$ be u-Cauchy and $\mathfrak{F} \to 0$ under v. There is a point $\tilde{x} \in \tilde{E}_u$ such that $\mathfrak{F}$ is the base of a filter in $\tilde{E}_u$ converging to $\tilde{x}$. Since $\langle \tilde{E}_u,E'\rangle$ is also a dual pair, for each $x' \in E'$, $\langle \mathfrak{F},x'\rangle$ is the base of a filter in $\mathbb{R}$ converging to $\langle \tilde{x},x'\rangle$. But $\mathfrak{F} \to 0$ under v and x' is v-continuous and so $\langle \mathfrak{F},x'\rangle \to 0$. Hence $\langle \tilde{x},x'\rangle = 0$ for all $x' \in E'$ and so $\tilde{x} = 0$. Thus $\mathfrak{F} \to 0$ under u and so the filter condition holds.

Lemma 3. Let E and F be t.v. spaces and f a linear continuous map of E into F. Let $\tilde{f}$ denote the continuous extension of f, which maps $\tilde{E}$ into $\tilde{F}$. Then $f^{-1}(0) = \tilde{f}^{-1}(0)$ iff the filter condition holds.

Proof: Assume that the filter condition holds. Since $f(x) = \tilde{f}(x)$ for $x \in E$, $f^{-1}(0) \subseteq \tilde{f}^{-1}(0)$. To show that $\tilde{f}^{-1}(0) \subseteq f^{-1}(0)$, let $\tilde{x} \in \tilde{E}$ such that $\tilde{f}(\tilde{x}) = 0$. Let $\mathfrak{F}$ be a Cauchy filter on E which is the base of a filter $\mathfrak{F}'$ on $\tilde{E}$ converging to $\tilde{x}$. Since $\tilde{f}$ is continuous, $\tilde{f}(\mathfrak{F}') \to \tilde{f}(\tilde{x}) = 0$. For each $B \in \mathfrak{F}'$, there is a subset A of B such that $A \in \mathfrak{F}$ and $f(A) = \tilde{f}(A) \subseteq \tilde{f}(B) \cap f(E)$, and so $f(\mathfrak{F})$ is finer than the trace on $f(E)$ of $\tilde{f}(\mathfrak{F}')$. Hence $f(\mathfrak{F}) \to 0$. By the filter condition it follows that $\mathfrak{F}$ converges to a point of E. Hence $\tilde{x} \in E$ and $f(\tilde{x}) = 0$, i.e., $\tilde{f}^{-1}(0) \subseteq f^{-1}(0)$.

Thus $f^{-1}(0) = \tilde{f}^{-1}(0)$. For the converse, let $\mathfrak{F}$ be a Cauchy filter on E with $f(\mathfrak{F}) \to f(x)$. Then there is a point $\tilde{x} \in \tilde{E}$ and a filter $\mathfrak{F}'$ on $\tilde{E}$, with base $\mathfrak{F}$ converging to $\tilde{x}$. Hence $\tilde{f}(\mathfrak{F}') \to \tilde{f}(x)$. Now if $A \in \mathfrak{F}$, then $f(A) \in \tilde{f}(\mathfrak{F}')$ and so $f(\mathfrak{F})$ is coarser than the trace on $f(E)$ of $\tilde{f}(\mathfrak{F}')$. Hence $\tilde{f}(\tilde{x}) = f(x)$ because $f(\mathfrak{F}) \to f(x)$. Thus $\tilde{x}-x \in \tilde{f}^{-1}(0) = f^{-1}(0)$ and so $\tilde{x} \in E$. This proves that $\mathfrak{F}$ is convergent to a point of E.

<u>Theorem 3</u>. Let E be an ultrabarrelled space and F a metrizable t.v. space. If f is a continuous linear map of F onto E and if the filter condition holds, then f is open.

<u>Proof</u>: Note that f can be extended to a continuous map $\tilde{f}$ of $\tilde{F}$ into $\tilde{E}$. Now $\tilde{F}$ is complete and metrizable and $\tilde{f}(\tilde{F})$ is ultrabarrelled by Corollary 3. Hence by Proposition 6, $\tilde{f}$ is an open map of $\tilde{F}$ onto $\tilde{f}(\tilde{F})$. Let V be a closed neighbourhood of 0 in F. Then $\tilde{f}(\tilde{V})$ is a neighbourhood of 0 in $\tilde{f}(\tilde{F})$ and so $\tilde{f}(\tilde{V}) \cap E$ is a neighbourhood of 0 in E. We claim that $\tilde{f}(\tilde{V}) \cap E \subset f(V)$. Let $y \in \tilde{f}(\tilde{V}) \cap E$. Then there exists $\tilde{x} \in \tilde{V}$ such that $\tilde{f}(\tilde{x}) = y$, and $x \in F$ so that $f(x) = y$ because f is onto. Therefore $\tilde{f}(\tilde{x}) - f(x) = \tilde{f}(\tilde{x}-x) = 0$. By Lemma 3, it follows that $\tilde{x} - x \in F$. Thus $\tilde{x} \in \tilde{V} \cap F = V$ because V is closed in F and so $y = \tilde{f}(\tilde{x}) \in \tilde{f}(V) = f(V)$. Hence $f(V)$ is a neighbourhood of 0 in E and so f is open.

<u>Theorem 4</u>. Let F be a metrizable t.v. space and E an ultrabarrelled t.v. space. If g is a linear map of E into F with the closed graph and if the inverse filter condition holds,

then g is continuous.

Proof: Note that g can be regarded as a map of E into $\tilde{F}$. We show that the graph G of g is closed in $E \times \tilde{F}$. Let $(x,y) \in \bar{G}$. If β and η are neighbourhood bases at 0 in E and $\tilde{F}$ respectively, then $[(x+U) \times (y+V)] \cap G \neq \emptyset$ for each $U \in \beta$ and $V \in \eta$. Hence $(x+U) \cap g^{-1}(y+V) \neq \emptyset$. Put $A_{U,V} = (x+U) \cap g^{-1}(y+V)$.

Let $\mathfrak{F}$ be the filter generated by the sets $A_{U,V}$ as U and V run through β and η. Then $\mathfrak{F} \to x$ and $g(\mathfrak{F}) \to y$. Therefore $g(\mathfrak{F})$ is Cauchy and so by the inverse filter condition, $g(\mathfrak{F})$ converges to a point in $g(E)$. In other words, $y \in g(E) \subset F$. This shows that $(x,y) \in \bar{G} \cap (E \times F)$. But G is closed in $E \times F$ and so $(x,y) \in G$. Therefore, G is closed in $E \times \tilde{F}$. Hence, by Theorem 2, g is a continuous map of E into $\tilde{F}$. Since the map $F \to \tilde{F}$ is continuous, it follows that $g\colon E \to F$ is continuous.

We now show that ultrabarrelled spaces have the property that the Banach-Steinhaus theorem holds for continuous linear maps into any t.v. space.

Theorem 5. Let (E,u) be an ultrabarrelled (quasiultrabarrelled) space and (F,v) any t.v. space. If H is a set of continuous linear maps of E into F, such that it is pointwise bounded (uniformly bounded on bounded sets), then H is equicontinuous.

Proof: Let (E,u) be an ultrabarrelled space. Let V be a closed circled neighbourhood of 0 in F. Then it easily follows that $\bigcap_{f \in H} f^{-1}(V)$ is an ultrabarrel in (E,u) and hence a neighbourhood of 0. This proves that H is equicontinuous. The proof can similarly be adapted when (E,u) is quasiultrabarrelled.

Remark 6. Waelbroeck [54] has shown that if the conclusion of Theorem 5 holds, (E,u) is necessarily ultrabarrelled. We refer the reader to Chapter 9 where we have adopted his proof in a different situation.

If (E,u) is a t.v. space, we write $u^{\circ\circ}$ to denote the l.c. topology derived from u; i.e., $u^{\circ\circ}$ is the finest l.c. topology coarser than u.

Proposition 7. If (E,u) is an ultrabarrelled (quasiultrabarrelled) space, then $(E,u^{\circ\circ})$ is barrelled (quasibarrelled).

Proof: Let B be a barrel in $(E,u^{\circ\circ})$. Then B is an ultrabarrel in (E,u) and so a convex neighbourhood of 0 in (E,u). The set B must then be a neighbourhood of 0 in $(E,u^{\circ\circ})$ which proves that $(E,u^{\circ\circ})$ is a barrelled space.

Corollary 10. Any l.c. ultrabarrelled (quasiultrabarrelled) space is barrelled (quasibarrelled).

If (E,u) is a metrizable t.v. space then $(E,u^{\circ\circ})$ is metrizable and so, if $\langle E,E'\rangle$ is a dual pair, $u^{\circ\circ} = \tau(E,E')$ ([4], Chapter 4, §2, Proposition 6).

Proposition 8. If (E,u) is a metrizable t.v. space with dual E', $\langle \tilde{E}_u, E'\rangle$ is a dual pair and $(E,u^{\circ\circ})$ is ultrabarrelled, then $u = u^{\circ\circ}$.

Proof: By Lemma 2, the filter condition holds for the identity map i of (E,u) onto $(E,u^{\circ\circ})$. Hence i is open by Theorem 3, and this proves that $u = u^{\circ\circ}$.

Corollary 11. If (E,u) is metrizable and $(E,u^{\circ\circ})$ is complete, then $u = u^{\circ\circ}$.

We now give an example, presented by Robertson [44] and due to Weston, which shows that a barrelled space need not be ultrabarrelled.

Example 2. Consider $\ell^{\frac{1}{2}} = \{x = \{x_n\}: \sum_{n=1}^{\infty} |x_n|^{\frac{1}{2}} < \infty\}$ equipped with the complete metrizable topology u determined by the neighbourhood basis $\{V_n\}$ at 0, where

$$V_n = \{x \in \ell^{\frac{1}{2}}: \|x\|_{\frac{1}{2}} \leq \frac{1}{n}\},\ n \geq 1,$$

$$\|x\|_{\frac{1}{2}} = \left(\sum_{n=1}^{\infty} |x|^{\frac{1}{2}}\right)^2,\ x = \{x_n\} \in \ell^{\frac{1}{2}}.$$

Clearly (E,u) is ultrabarrelled and so $(E,u^{\circ\circ})$ is barrelled. If we write $\|x\|_1 = \sum_{n=1}^{\infty} |x_n|$, then since $\|x\|_1 \leq \|x\|_{\frac{1}{2}}$, $\ell^{\frac{1}{2}}$ is a subspace of ℓ^1 and the topology induced on $\ell^{\frac{1}{2}}$ by the norm topology of ℓ^1 is coarser than u. Since the space ϕ of all sequences with only finitely many nonzero components is dense in ℓ^1 under the norm topology and $\phi \subseteq \ell^{\frac{1}{2}}$, $\ell^{\frac{1}{2}}$ is dense in ℓ^1 under the norm topology. Hence the dual of $\ell^{\frac{1}{2}}$ under the norm topology is ℓ^{∞} and the norm topology is thus $\tau(\ell^{\frac{1}{2}},\ell^{\infty})$. Also, the dual of $\ell^{\frac{1}{2}}$ under u is ℓ^{∞}, and so $u^{\circ\circ} = \tau(\ell^{\frac{1}{2}},\ell^{\infty})$. In view of Proposition 8, we conclude that $(\ell^{\frac{1}{2}},u^{\infty})$ is not ultrabarrelled.

§2. Ultrabornological Spaces.

Definition 4. A t.v. space (E,u) is called ultrabornological if each bounded linear map of E into any t.v. space is continuous.

Remark 7. Clearly a metrizable t.v. space is ultrabornological.

A countable dimensional metrizable not necessarily locally convex t.v. space is ultrabornological. But it is not ultrabarrelled by Remark 3.

The following example shows that an ultrabarrelled space need not be ultrabornological.

Example 3. Let (E,u) be an incomplete (Hausdorff) inductive limit of a sequence of Banach spaces and $\tilde{E}$ the completion of E. Let $x \in \tilde{E} \setminus E$. Then the vector subspace E_1 of $\tilde{E}$ spanned by E and x is not bornological in view of a result of Y. Kōmura [29], page 155, and hence is not ultrabornological. But E_1 is ultrabornological by Corollary 9, and Proposition 14 of [44].

The proof of the following theorem is simple and is omitted.

Theorem 6. Let (E,u) be a t.v. space and η the family of all bornivorous suprabarrels in (E,u). Then η is a base of neighbourhoods of 0 for the finest vector topology v on E with the same bounded sets as u. (E,u) is ultrabornological iff u = v i.e. iff each bornivorous suprabarrel in (E,u) is a neighbourhood of 0.

Remark 8. v is called the ultrabornological topology associated with u.

Corollary 12. Every ultrabornological space is quasiultrabarrelled.

Analogous to Theorem 2, we have the following:

Theorem 7. Let (E,u) be a t.v. space. The following statements are equivalent:

(a) (E,u) is ultrabornological.

(b) Each bounded linear map of (E,u) into any complete metrizable t.v. space is continuous.

(c) Each F-seminorm p on E is continuous if $\sup\{p(sx): x \in B\} < \infty$ for $s \to 0$ and for all $B \in G$, where G is the family of all bounded sets.

Proposition 9. If (E,u) is ultrabornological then $(E,u^{\circ\circ})$ is bornological.

Proof: Let g be a bounded linear map of $(E,u^{\circ\circ})$ into any l.c. space F. Then g, being a bounded linear map of (E,u) into F, is continuous. Since F is a l.c. space, it follows that g is a continuous linear map of $(E,u^{\circ\circ})$ into F. Thus $(E,u^{\circ\circ})$ is bornological.

Corollary 13. Any l.c. ultrabornological space is bornological.

That a bornological space need not be ultrabornological is illustrated in the following:

Example 4. Let E be an uncountable dimensional topological vector space. Then its finest vector topology w is strictly finer than $\tau(E,E^*)$ ([44], Page 299) so that the identity map of $(E,\tau(E,E^*))$ into (E,w) is not continuous. But it is bounded, since every $\tau(E,E^*)$-bounded subset is contained in a finite-dimensional subspace of E. Hence $(E,\tau(E,E^*))$ is not ultrabornological, though it is bornological.

Proposition 10. Any *-inductive limit of ultrabornological spaces is ultrabornological.

Proof: Let E be the *-inductive limit of a family of ultrabornological spaces $\{[E_\alpha: f_\alpha]; \alpha \in A\}$. Let g be a bounded linear map of E into a t.v. space F. Clearly $g \circ f_\alpha$ is a bounded linear map of E_α into F. Since E_α is ultrabornological, $g \circ f_\alpha$ is continuous for all α. Hence g is continuous which implies that E is ultrabornological.

As an easy consequence, we have:

Corollary 14. Any countable inductive limit of l.c. ultrabornological spaces is ultrabornological.

Remark 9. It follows from Corollary 14 and Kōmura's example of a non-bornological barrelled space ([29], p. 155) that a nonultrabornological space may contain a dense ultrabornological space.

Proposition 11. (Adasch [1]). The topological product of a

countable family of ultrabornological spaces is ultrabornological.

Proof: Let (E_i, u_i) be a countable family of ultrabornological spaces and $(E,u) = \prod_{i=1}^{\infty} (E_i, u_i)$. Let V be a bornivorous suprabarrel in (E,u) with $\{V_i\}$ as a defining sequence. First we show that

(a) there exists an n_0 such that $V_1 \supset \prod_{n_0}^{\infty} E_i$

Assume that it is false.

Then there exists $x^{(1)} \in \prod_{i=2}^{\infty} E_i$ with $x^{(1)} \notin V_1$. Further, there exists $x^{(2)} \in \prod_{i=3}^{\infty} E_i$ with $x^{(2)} \notin 2V_1$.

Continuing in this manner, we obtain the sequence

$$x^{(n)} \in \prod_{n+1}^{\infty} E_i \quad \text{with } x^{(n)} \notin nV_1.$$

It is easy to show that the set consisting of the $x^{(n)}$ is bounded in $\prod_1^{\infty} E_i$. But, since the suprabarrel V_1 does not absorb this set, (a) holds. The rest of the proof is similar to the proof of Proposition 4' (part (b_1)).

Remark. The proof of Proposition 18 in Chapter 2 is a simple adaptation of that of Proposition 11.

CHAPTER IV

ORDER-QUASIBARRELLED VECTOR LATTICES

The concept of Order-infrabarrelled Riesz space, which we shall henceforth call order-quasibarrelled vector lattice, is due to Wong [61]. Most of the results of this Chapter are due to him. Section 10, Chapter 1 may be consulted for elementary facts about lattices.

§1. Order-quasibarrelled vector lattices.

Definition 1. An l.c. vector lattice (E,C,u) is called order-quasibarrelled if each order-bornivorous barrel in (E,C,u) is a neighbourhood of 0.

Let (E,C) be a vector lattice and B a subset of (E,C). The set K(B) defined by

$$K(B) = \{x \in E: [-|x|, |x|] \subseteq B\}$$

is said to be the solid kernel of B. It is the largest solid set contained in B and it is convex if B is convex.

Lemma 1. Let V be a circled subset of a vector lattice (E,C). Then K(V) is order-bornivorous iff V is so.

Proof: Since K(V) $\subseteq$ V, the "only if" part is obvious. For the "if" part, we proceed as follows: Suppose K(V) is not order-bornivorous. Then there exists t ε C such that

$[-t,t] \not\subseteq n\,K(V)$ for each natural number n. For each n, let $x_n \in [-t,t] \setminus n\,K(V)$. Then there exists y_n such that $y_n \in [-|n^{-1}x_n|, |n^{-1}x_n|]$ and $y_n \notin V$. This implies that $n\,y_n \in [-|x_n|, |x_n|] \subseteq [-t,t]$ and $n\,y_n \notin n\,V$ for all n. Hence V is not order-bornivorous which is a contradiction.

<u>Theorem 1</u>. An l.c. vector lattice (E,C,u) is order-quasibarrelled iff each solid barrel in (E,C,u) is a neighbourhood of 0.

<u>Proof</u>: Assume that (E,C,u) is order-quasibarrelled. Let V be a solid barrel in (E,C,u). Since V is solid and absorbing, it follows that V is order-bornivorous and hence a neighbourhood of 0. Conversely assume that the condition is satisfied. Let U be an order-bornivorous barrel in (E,C,u) and K(U) the solid kernel of U. K(U) is order-bornivorous by Lemma 1, and this implies that K(U) is absorbing. Since K(U) is the largest solid subset of U, it follows from

$$K(U) \subseteq \overline{K(U)} \subset \overline{U} = U$$

and (Chapter 1, Proposition 33(a)) that K(U) is closed. Clearly K(U) is convex. Thus K(U) is a solid barrel and hence a neighbourhood of 0 in E. But $K(U) \subseteq U$ and hence U is a neighbourhood of 0. This completes the proof.

Let (E,C,u) be an l.c. vector lattice. For any given $x \in E$, let $p_x(f) = |f|(x) = \sup_{y \in [-x,x]} |\langle y,f\rangle|$ for all $f \in E'$. Then $\{p_x;\ x \in C\}$ defines a locally solid topology on E', denoted by $\sigma_s(E',E)$, such that $(E',C',\sigma_s(E',E))'$ is the lattice ideal in

E'^b generated by E. This topology $\sigma_s(E',E)$ is called the locally solid topology associated with $\sigma(E',E)$. Similarly we have the topology $\sigma_s(E,E')$ on E. It is easily seen that $\sigma(E',E)$ is coarser than $\sigma_s(E',E)$ and that $\sigma_s(E',E)$ is coarser than $\beta(E',E)$ on E'.

Theorem 2. Let (E,C,u) be an l.c. vector lattice. The following statements are equivalent:

(a) (E,C,u) is order-quasibarrelled.

(b) Each order-bounded lower semicontinuous seminorm on E is continuous.

(c) Each lower semicontinuous lattice seminorm on E is u-continuous.

(d) Each $\sigma_s(E',E)$-bounded subset of E' is equicontinuous.

Proof: A seminorm p on E is order-bounded and lower semicontinuous iff the set $V = \{x \in E;\ p(x) \leq 1\}$ is an order-bornivorous barrel in (E,C,u) and hence (a) and (b) are equivalent. A seminorm q on E is a lower semicontinuous lattice seminorm iff the set $U = \{x \in E;\ q(x) \leq 1\}$ is a solid barrel in (E,C,u) and hence (a) and (c) are equivalent. A subset W of E is an order-bornivorous barrel in (E,C,u) iff W^0 is a $\sigma_s(E',E)$-bounded subset of E' and hence (a) and (d) are equivalent.

The following corollary can easily be verified and hence its proof is omitted.

Corollary 1. Every barrelled l.c. vector lattice is order-quasibarrelled and every order-quasibarrelled vector lattice is quasibarrelled. Further, a vector lattice equipped with the order-bound topology is order-quasibarrelled.

The following examples show that an order-quasibarrelled vector lattice need not be barrelled and that a quasibarrelled l.c. vector lattice need not be order-quasibarrelled.

<u>Example 1</u>. Let $m = \ell^{\infty}$ be the Banach lattice of all bounded real sequences with the usual pointwise ordering and the supremum norm $\|.\|$. For $n \geq 1$ let e_n be the sequence having 1 in the nth component and 0 elsewhere. Let E_0 be the subspace of m generated by the $\{e_n\}$ so that E_0 consists of all finite sequences. Let e be the sequence having 1 in every coordinate, and let

$$E = \{x + \lambda e: \ x \in E_0, \lambda \in \mathbb{R}\},$$

with the supremum norm $\|.\|$ induced from m. Then E is an order-unit-normed vector lattice with e as order-unit. It is easy to see that an order-unit-norm topology is an order-bound topology and that each vector lattice with an order-bound topology is always order-quasibarrelled. Thus E is order-quasibarrelled. We now show that it is not barrelled. For $n \geq 1$, define the continuous linear functional f_n on E by $f_n(x + \lambda e) = n\, x_n$, $x = \{x_n\}$. It is well-defined, and the set $\{f_n\}$ is pointwise bounded; however $\{f_n\}$ is not equicontinuous so that E is not barrelled.

<u>Example 2</u>. Let ϕ be the normed vector lattice of all real

sequences having only finitely many nonzero components, equipped with the usual pointwise ordering and the supremum-norm $\|.\|$. Clearly ϕ is quasibarrelled. We now show that it is not order-quasibarrelled. Define $V = \{x = (x_n) \in \phi;\ |x_n| \leq \frac{1}{n}$ for all $n \geq 1\}$. Then V is a solid barrel in ϕ which is not a neighbourhood of 0. Hence ϕ is not order-quasibarrelled.

Now we consider various conditions under which an order-quasibarrelled vector lattice is barrelled and a quasi-barrelled l.c. Riesz space is order-quasibarrelled.

<u>Lemma 2</u>. If (E,C,u) is an order-quasibarrelled vector lattice, then E' is a normal subspace of E^b.

<u>Proof:</u> Let $0 \leq f \in E'$ and $0 \leq f_\alpha \uparrow f$ in E^b. Then $\sup f_\alpha(t) = f(t)$ for any $t \in C$. It follows that $\{f_\alpha\}$ is $\sigma_s(E',E)$-bounded and hence equicontinous because E is order-quasibarrelled. On the the other hand, we show that f_α converges to f pointwise. For any $x \in E$ and $\delta > 0$, there exists α_0 in D such that

$$f(|x|) - \delta < f_{\alpha_0}(|x|) \leq f(|x|).$$

Now it follows from $f_\alpha \uparrow f$ that $|f_\alpha(|x|) - f(|x|)| < \delta$, whenever $\alpha \geq \alpha_0$. We therefore have

$$|f_\alpha(x) - f(x)| \leq |f_\alpha - f|(|x|) = (f - f_\alpha)(|x|) < \delta,$$

whenever $\alpha \geq \alpha_0$. This shows that f_α converges to f pointwise on E. Since $\{f_\alpha\}$ is equicontinuous we conclude that $f \in E'$ and hence E' is a normal subspace of E^b.

Corollary 2. Let (E,C,u) be an order-quasibarrelled vector lattice. Then $(E',\beta(E',E))$ is complete.

Proof: It is known ([47], page 237) that $(E',\beta(E',E))$ is complete whenever E' is a normal subspace of E^b. Hence the result follows by Lemma 2.

Lemma 3. Let (E,C,u) be an l.c. vector lattice and u'_b the order-bound topology on E'. If E' is a normal subspace of E^b, then a subset B of E' is $\sigma_s(E',E)$-bounded iff B is u'_b-bounded.

Proof: The "if"part is clear because u'_b is finer than $\sigma_s(E',E)$. To prove the "only if" part we assume, without loss of generality, that B, in addition, is solid and convex. Assume that B is not u'_b-bounded. Then there exists a solid and convex neighbourhood V of 0 in (E',u'_b) such that $B \nsubseteq 2^{2n}V$ for each natural number n. For each n, there exists $f_n \in E'$ such that $|f_n| \in B$ and $|f_n| \notin 2^{2n}V$. Let $g_n = \sum_{k=1}^{n} 2^{-k}|f_k|$. Then $\{g_n\}$ is a $\sigma_s(E',E)$-Cauchy sequence in E', because B is convex and $\sigma_s(E',E)$-bounded. Since E' is a normal subspace of E^b, it follows from (Chapter 1, Proposition 43) that $(E',\sigma_s(E',E))$ is complete. Hence there exists $g \in E'$ such that $g_n \to g$ in $\sigma_s(E',E)$-topology. Since $0 \leq g_1 \leq g_2 \leq \ldots$ and C' is $\sigma_s(E',E)$-closed, we conclude that $0 \leq g_n \leq g$ for all n. Since V is a neighbourhood of 0 in (E',u'_b), there exists k such that $[0,g] \subseteq 2^k V$. It follows from $0 \leq 2^{-k}|f_k| \leq g_k \leq g$ that $|f_k| \in 2^{2k}V$ which is a contradiction. Hence B is u'_b-bounded.

Lemma 4. Let (E,C,u) be an order-quasibarrelled vector lattice. Then C' is a strict $\mathfrak{B}$-cone in $(E',C',\sigma(E',E))$ iff it is a $\mathfrak{B}$-cone in $(E',C',\sigma(E',E))$.

Proof: The "if" part is clear. To prove the "only if" part let B be any $\sigma(E',E)$-bounded subset of E'. There exists a $\sigma(E',E)$-bounded subset A of E' such that

$$B \subset \overline{A \cap C' - A \cap C'},$$

closure being taken in $\sigma(E',E)$-topology. Let H be the $\sigma(E',E)$-closure of A. Then H is a $\sigma(E',E)$-bounded subset of E' such that the $\sigma(E',E)$-closure of $A \cap C' - A \cap C'$ is contained in the $\sigma(E',E)$-closure of $H \cap C' - H \cap C'$. We claim that $H \cap C' - H \cap C'$ is $\sigma(E',E)$-closed. Since $H \cap C'$ is a positive $\sigma(E',E)$-bounded subset of E', it is $\sigma_s(E',E)$-bounded. Hence $H \cap C'$ is an equicontinuous $\sigma(E',E)$-closed subset of E' and so $\sigma(E',E)$-compact in view of the Alaoglu-Bourbaki Theorem. Consequently $H \cap C' - H \cap C'$ is $\sigma(E',E)$-closed. Thus B is contained in $H \cap C' - H \cap C'$, which shows that C' is a strict $\mathfrak{B}$-cone in $(E',C',\sigma(E',E))$.

Lemma 5. Let (E,C,u) be an l.c. vector lattice and let $E''_{|\sigma|} = (E',\sigma_s(E',E))'$. Then the following statements are equivalent:

(a) Each $\sigma_s(E',E)$-bounded set in E' is $\beta(E',E)$-bounded.

(b) The topology $\beta(E''_{|\sigma|},E')$ on $E''_{|\sigma|}$ is the relative topology induced by $\beta(E'',E')$ and the $\sigma_s(E',E)$-closure of each $\beta(E',E)$-bounded subset of E' is $\beta(E',E)$-bounded, where $E'' = (E',\beta(E',E))'$.

Proof: (a)$\Rightarrow$(b): Let V" be any neighbourhood of 0 in $(E''_{|\sigma|}, \beta(E''_{|\sigma|}, E'))$. There exists a $\sigma_s(E',E)$-bounded subset B of E' such that the polar B^o of B with respect to $\langle E', E''_{|\sigma|}\rangle$ is contained in V". By assumption B is $\beta(E',E)$-bounded and the polar B^o of B with respect to $\langle E', E''\rangle$ is a neighbourhood of 0 in $(E'', \beta(E'', E'))$. Since $B^o(E''_{|\sigma|}) = B^o(E'') \cap E''_{|\sigma|}$, we have $B^o(E'') \cap E''_{|\sigma|} \subseteq V''$. Hence V" is a neighbourhood of 0 in $E''_{|\sigma|}$ induced by $\beta(E'',E')$. Clearly the relative topology on $E''_{|\sigma|}$ induced by $\beta(E'',E')$ is always coarser than $\beta(E''_{|\sigma|}, E')$. Hence $\beta(E''_{|\sigma|}, E')$ is the relative topology on $E''_{|\sigma|}$ induced by $\beta(E'',E')$. The other part is obvious.

(b)$\Rightarrow$(a): Let B be any $\sigma_s(E',E)$-bounded set in E'. Then the polar B^o of B with respect to $\langle E', E''_{|\sigma|}\rangle$ is a neighbourhood of 0 in $(E''_{|\sigma|}, \beta(E''_{|\sigma|}, E'))$. There exists a circled, convex and bounded subset A of $(E', \beta(E',E))$ such that $A^o(E'') \cap E''_{|\sigma|} \subseteq B^o(E''_{|\sigma|})$. Let V be the $\sigma_s(E',E)$-closure of A. We conclude from $A^o(E''_{|\sigma|}) = A^o(E'') \cap E''_{|\sigma|}$ and $V^o(E''_{|\sigma|}) \subseteq A^o(E''_{|\sigma|})$ that $V^o(E''_{|\sigma|}) \subseteq B^o(E''_{|\sigma|})$ and hence that $V = [V^o(E''_{|\sigma|})]^o \cap (E') \supseteq B$. This shows that B is $\beta(E',E)$-bounded.

Lemma 6. Let (E,C,u) be an l.c. vector lattice and let $E''_{|\sigma|} = (E', C', \sigma_s(E',E))'$. Then the following statements are equivalent:

(a) Each $\sigma(E',E)$-bounded set in E' is $\sigma_s(E',E)$-bounded.

(b) The topology $\beta(E,E')$ on E is the relative topology induced by $\beta(E''_{|\sigma|}, E')$ and the $\sigma(E',E)$-closure of each $\sigma_s(E',E)$-bounded set in E' is $\sigma_s(E',E)$-bounded.

Proof: This is similar to that of Lemma 5.

Proposition 1. Let (E,C,u) be a quasibarrelled l.c. vector lattice. Then it is order-quasibarrelled iff E' is a normal subspace of E^b.

Proof: The "only if" part follows from Lemma 2. To prove the "if" part, we observe that $\beta(E',E)$ is coarser than the order-bound topology u'_b on E'. It follows from Lemma 3 that each $\sigma_s(E',E)$-bounded subset of E' is $\beta(E',E)$-bounded. Hence (E,C,u) is order-quasibarrelled by Theorem 2.

Corollary 3. Let (E,C,u) be a bornological l.c. vector lattice. Then it is order-quasibarrelled iff E' is a normal subspace of E^b.

Proposition 2. Let (E,C,u) be a quasibarrelled l.c. vector lattice. Then it is order-quasibarrelled iff the topology $\beta(E''_{|\sigma|},E')$ on $E''_{|\sigma|}$ is the relative topology induced by $\beta(E'',E')$ and the $\sigma_s(E',E)$-closure of each $\beta(E',E)$-bounded set in E' is $\beta(E',E)$-bounded.

Proof: This follows from Theorem 2(d) and Lemma 5.

Lemma 7. Let (E,C,u) be a vector lattice equipped with an l.c. topology u such that the cone C is normal for u. Let $\mathcal{B}$ be a neighbourhood basis at 0 for u consisting of circled, convex and full sets in E. Let $\mathcal{B}_s = \{K(V);\ V \in \mathcal{B}\}$. Then there exists a unique locally solid topology u_s on E such that $\mathcal{B}_s$ is a neighbourhood basis at 0 for u_s. Furthermore, u_s is finer than u, and u_s is the greatest lower bound of all locally solid topologies which are finer than u.

Proof: Clearly each K(V) is non-empty, convex, solid and absorbing. Furthermore, $K(\lambda V) = \lambda K(V)$ for all $\lambda > 0$, $K(V_1) \cap K(V_2) = K(V_1 \cap V_2)$, and $K(V) \subseteq K(U)$ whenever $V \subseteq U$. Hence there exists a unique locally solid topology u_s on E such that $\mathcal{B}_s$ is a neighbourhood basis at 0 for u_s. Since $K(V) \subseteq V$, it follows that u_s is finer than the Hausdorff topology u and so u_s is Hausdorff. Now let u' be any locally solid topology on E such that u' is finer than u and let W be any u_s-neighbourhood of 0. There exists $V \in \mathcal{B}$ such that $K(V) \subseteq W$. Since u is coarser than u', it follows that $U \subseteq V$ for some convex and solid u'-neighbourhood U of 0 and therefore $U \subseteq K(V)$. This shows that u' is finer than u_s.

Definition 2. u_s as introduced in Lemma 7 is called the locally solid topology on E associated with u.

Remark 1. (1) u is locally solid topology on E iff $u = u_s$.
(2) If u is metrizable (normable), so is u_s.

Lemma 8. Let (E,C) and u be as in Lemma 7 and u_s as in Definition 2. Then a positive subset B of E is u_s-bounded iff it is u-bounded.

Proof: The "only if" part is obvious. To prove the "if" part we suppose that B is u-bounded but not u_s-bounded. Then there exists a circled, convex and full u-neighbourhood V of 0 such that $B \not\subseteq n\,K(V)$ for each natural number n. Since $K(V) \cap C = V \cap C$, $B \not\subseteq n(V \cap C)$ and so $B \not\subseteq n\,V$. But this implies that B is not u-bounded which is a contradiction. This proves the lemma.

<u>Lemma 9</u>. Let (E,C), u and u_s be as in Lemma 8. Then each u-bounded subset of E is u_s-bounded iff the positive cone C is a strict $\mathcal{B}_u$-cone.

<u>Proof</u>: Since C is a strict $\mathcal{B}_{u_s}$-cone, it follows that C is a strict $\mathcal{B}_u$-cone, provided that each u-bounded set is u_s-bounded. Conversely, if B is any u-bounded subset of E, there exists an u-bounded subset A of E such that $B \subseteq A \cap C - A \cap C$. It follows from Lemma 8 that $A \cap C$ is u_s-bounded. Therefore $A \cap C - A \cap C$ is u_s-bounded, in particular, B is u_s-bounded.

<u>Proposition 3</u>. Let (E,C,u) be an order-quasibarrelled vector lattice. Then it is barrelled iff C' is a $\mathcal{B}$-cone in $(E',C',\sigma(E',E))$.

<u>Proof</u>: Assume that (E,C,u) is barrelled. Then each subset B of E' is $\sigma(E',E)$-bounded iff it is $\sigma_s(E',E)$-bounded. Since C' is a strict $\mathcal{B}$-cone in $(E',C',\sigma_s(E',E))$, it follows that C' is a strict $\mathcal{B}$-cone in $(E',C',\sigma(E',E))$; in particular C' is a $\mathcal{B}$-cone in $(E',C',\sigma(E',E))$. Conversely, assume that C' is a $\mathcal{B}$-cone in $(E',C',\sigma(E',E))$. By Lemma 4, C' is a strict

$\mathfrak{B}$-cone in $(E',C',\sigma(E',E))$. It now follows from Lemma 9 that each $\sigma(E',E)$-bounded subset of E' is $\sigma_s(E',E)$-bounded, which implies that (E,C,u) is barrelled by Theorem 2.

Corollary 4. Let (E,C,u) be a quasibarrelled l.c. vector lattice. Then (E,C,u) is barrelled iff E' is a normal subspace of E^b and C' is a $\mathfrak{B}$-cone in $(E',C',\sigma(E',E))$.

Proof: The "only if part" follows from Lemma 2 and Proposition 3. The "if part" follows from Proposition 1 and 3.

Corollary 5. Let (E,C,u) be a bornological l.c. vector lattice. Then (E,C,u) is barrelled iff E' is a normal subspace of E^b and C' is a $\mathfrak{B}$-cone in $(E',C',\sigma(E',E))$.

Proposition 4. Let (E,C,u) be an order-quasibarrelled vector lattice. Then it is barrelled iff the topology $\beta(E,E')$ on E is the relative topology induced by $\beta(E''_{|\sigma|},E')$ and the $\sigma(E',E)$-closure of each $\sigma_s(E',E)$-bounded set in E' is $\sigma_s(E',E)$-bounded.

Proof: This follows from Lemma 6.

Let (E,C,u) be an l.c. vector lattice, and let $t \in C$. Then $E_t = \bigcup_n n\,[-t,t]$ is the lattice ideal in E generated by t. If p_t denotes the gauge of $[-t,t]$ on E_t and $C_t = C \cap E_t$, then (E_t,C_t,p_t) is a normed vector lattice for which the relative topology on E_u induced by u is coarser than the norm topology on E_u given by p_t.

<u>Proposition 5</u>. Let (E,C,u) be an order-quasibarrelled vector lattice. For any $t \in C$, if E_t is complete for the norm p_t, then (E,C,u) is barrelled.

<u>Proof</u>: Let V be any barrel in (E,C,u) and $t \in C$. Since E_t is complete for the norm p_t, (E_t, C_t, p_t) is barrelled. Clearly $V \cap E_t$ is a barrel in (E_t, C_t, p_t), because the relative topology on E_t is coarser than the norm topology given by p_t; consequently, $V \cap E_t$ absorbs $[-t,t]$. Since $\{[-t,t];\ t \in C\}$ forms a fundamental system of order-bounded sets in E, it follows that V is an order-bornivorous barrel in (E,C,u) and hence a neighbourhood of 0. This proves that (E,C,u) is barrelled.

<u>Corollary 6</u>. Each σ-Dedekind complete and order-quasibarrelled vector lattice (E,C,u) is barrelled.

<u>Proof</u>: Let $t \in C$. We wish to show that E_t is complete for the norm p_t; for this it is sufficient to show that if $0 \leq t_n \in E_n$ and if $\sum_{n=1}^{\infty} p_t(t_n) < \infty$, then $\sup_n \{ \sum_{i=1}^{n} t_i \}$ exists in E_t. Clearly $\{ \sum_{i=1}^{n} t_i;\ n \geq 1\}$ is a p_t-Cauchy sequence in E_t, and hence $\{ \sum_{i=1}^{n} t_i;\ n \geq 1\}$ is p_t-bounded. Therefore, there exists $\alpha > 0$ such that $p_t(\sum_{i=1}^{n} t_i) \leq \alpha$ for all n. It follows that $\sum_{i=1}^{n} t_i$ belongs to $\alpha[-t,t]$ for all n. Hence $\sup_n \{ \sum_{i=1}^{n} t_i \}$ exists in E_t because E_t is a lattice ideal in E. Now Proposition 5 applies to give the required result.

Proposition 6. Every sequentially complete order-quasibarrelled vector lattice is a barrelled l.c. vector lattice.

Proof: By Corollary 1, every order-quasibarrelled vector lattice is quasibarrelled. The result now follows from (Chapter 2, Proposition 5).

Corollary 7. Every quasicomplete (in particular, complete) order-quasibarrelled vector lattice is a barrelled l.c. vector lattice.

Proof: This follows from Corollary 1.

Let X be a completely regular Hausdorff space and let C(X)′ denote the vector space of all continuous real-valued functions on X equipped with the compact-open topology u_c, and ordered by the cone K, where $K = \{f \in C(X);\ f(x) \geq 0$ for all $x \in X\}$. Then $(C(X), K, u_c)$ is an l.c. vector lattice. A subset B of X is said to be C(X)-pseudocompact if each function in C(X) is bounded on B. Let $\phi \in C(X)'$. Then the support of ϕ, denoted by supp(ϕ), is the smallest closed subset A of X such that $\phi(f) = 0$, for every $f \in C(X)$ vanishing on A. Supp(f) always exists and is compact. Let F be a subset of C(X)′. Then the support of F, written Supp(F), is defined to be the closure of $\cup\{\text{supp}(\phi);\ \phi \in F\}$.

Proposition 8. Let X be a completely regular Hausdorff space. Then the following statements are equivalent:

(a) $(C(X),K,u_c)$ is barrelled.

(b) $(C(X),K,u_c)$ is order-quasibarrelled.

(c) Each closed $C(X)$-pseudocompact subset A of X is compact.

Proof: That (a) $\Rightarrow$ (b) follows from Corollary 1. The equivalence of (a) and (c) has been established in ([38],[49]). We show that (b) implies (c) to complete the proof. We observe that X is homeomorphic with some subset $\hat{X}$ of $(C(X)',\sigma(C(X)',C(X)))$ under the map $x \to \hat{x}$ defined by $\hat{x}(f) = f(x)$ for all $f \in C(X)$. It follows from $|\hat{x}|(|f|) = \sup\{|x(g)|; |g| \leq |f|, g \in C(X)\}$ that $\hat{A} = \{\hat{x}; x \in A\}$ is a $\sigma_s(C(X)',C(X))$-bounded subset of $C(X)'$, because A is $C(X)$-pseudocompact. Therefore, by Theorem 2 $\hat{A}$ is equicontinuous. It follows from (Chapter 5, Lemma 2) that $\mathrm{supp}(\hat{A})$ is a compact subset of X. Clearly, $A \subseteq \mathrm{supp}(\hat{A})$ and hence A is compact.

§2. Permanence Properties.

Let $\{(E_\alpha,C_\alpha); \alpha \in A\}$ be a family of vector lattices and let u_α be an l.c. topology on E_α such that C_α is normal for u_α. Let $E = \bigoplus_\alpha E_\alpha$ (the algebraic direct sum of the E_α's), $C = \bigoplus_\alpha C_\alpha$ and $u = \bigoplus_\alpha u_\alpha$ (the l.c. direct sum topology on E). Let $i_\alpha: E_\alpha \to E$ be the injection map and $\pi_\alpha: E \to E_\alpha$ the projection map, for each $\alpha \in A$.

Lemma 10. Let $u_{\alpha,s}$ denote the locally solid topology on E_α

associated with u_α for each α and let u_s be the locally solid topology on E associated with u. Then u_s is coarser than $\oplus u_{\alpha,s}$.

Proof: In view of (Chapter 1, Proposition 36(a)), $\oplus_\alpha u_\alpha$ is an l.c. topology on E for which C is normal. Let U be any u_s-neighbourhood of 0 in E. We show that $i_\alpha^{-1}(U)$ is an $u_{\alpha,s}$-neighbourhood of 0 in E_α for each $\alpha \in A$. Let W be a circled, convex and full neighbourhood of 0 in $(E,C,\oplus u_\alpha)$ such that $K(W) \subseteq U$. For each $\alpha \in A$, $i_\alpha^{-1}(W)$ is a circled, convex and full neighbourhood of 0 in $(E_\alpha,C_\alpha,u_\alpha)$ and so $K(i_\alpha^{-1}(W))$ is a neighbourhood of 0 in $(E_\alpha,C_\alpha,u_{\alpha,s})$. We now show that $K(i_\alpha^{-1}(W)) = i_\alpha^{-1}(K(W))$. Since $K(W) \subseteq W$, and since $i_\alpha^{-1}(K(W))$ is solid, it follows that $i_\alpha^{-1}(K(W)) \subseteq K(i_\alpha^{-1}(W))$. Conversely, suppose that $x_\alpha \in K(i_\alpha^{-1}(W))$. Then $i_\alpha(x_\alpha) \in W$. If $y \in E$ is such that $|y| \leq |i_\alpha(x_\alpha)|$, then $y = i_\alpha(\pi_\alpha(y))$ and $|\pi_\alpha(y)| \leq |x_\alpha|$. It follows that $\pi_\alpha(y) \in i_\alpha^{-1}(W)$, and so $y = i_\alpha(\pi_\alpha(y)) \in W$. We conclude that $i_\alpha(x_\alpha) \in K(W)$ or, equivalently, $x_\alpha \in i_\alpha^{-1}(K(W))$. This proves that $K(i_\alpha^{-1}(W)) = i_\alpha^{-1}(K(W))$. It now follows from $K(i_\alpha^{-1}(W)) = i_\alpha^{-1}(K(W)) \subseteq i_\alpha^{-1}(U)$ for each $\alpha \in A$ that U is a neighbourhood of 0 in $(E,C,\oplus u_{\alpha,s})$. Hence u_s is coarser than $\oplus u_{\alpha,s}$.

Corollary 8. Let u_α be a locally solid topology on E_α for each $\alpha \in A$. Then the l.c. direct sum topology $\oplus u_\alpha$ on E_α is a locally solid topology and hence $(E,C,\oplus u_\alpha)$ is a l.c. vector lattice.

Theorem 3. The l.c. direct sum of a family of order-quasibarrelled vector lattices is an order-quasibarrelled vector lattice.

Proof: Let $\{(E_\alpha, C_\alpha, u_\alpha);\ \alpha \in A\}$ be a family of order-quasibarrelled vector lattices. Let $E = \oplus E_\alpha$, $C = \pi_\alpha C_\alpha$, $K = C \cap E$, and $u = \oplus u_\alpha$ (the l.c. direct sum topology on E). Let $i_\alpha : E_\alpha \to E$ be the injection map for each α. Then (E,K,u) is an l.c. vector lattice (by Corollary 8). We now show that it is order-quasibarrelled. Let V be any solid barrel in (E,K,u). Since each i_α is a continuous lattice homomorphism of E_α into E, it follows that $i_\alpha^{-1}(V)$ is a solid barrel in $(E_\alpha, C_\alpha, u_\alpha)$ and hence a neighbourhood of 0 in $(E_\alpha, C_\alpha, u_\alpha)$. This implies that V is a neighbourhood of 0 in E. This shows that (E,K,u) is order-quasibarrelled.

The following result is due to N. Adasch [1].

Proposition 9. The topological product of a family of order-quasibarrelled vector lattices is order-quasibarrelled.

Proof: Let $\{(E_\alpha, C_\alpha, u_\alpha);\ \alpha \in A\}$ be a family of order-quasibarrelled vector lattices. The topological product (E,u) = $\pi_\alpha (E_\alpha, u_\alpha)$ ordered by the cone $C = \pi_\alpha C_\alpha$ is an l.c. vector lattice. Now we show that it is order-quasibarrelled. Let V be a solid barrel in E. Then it follows from (a) in the proof of Proposition 4', Chapter 3, that $\frac{1}{2}V \supset \pi_{A \setminus J} E_\alpha$ where J is a finite subset of A. The set $\frac{1}{2}V \cap \pi_\alpha E_\alpha$ is a solid barrel in $\pi_J E_\alpha$, and hence a neighbourhood of 0, since $\pi_J E_\alpha$ is order-quasibarrelled by Theorem 3. Hence it follows as in (b_1) in the proof of Proposition 4, Chapter 3 that $V = \frac{1}{2}V + \frac{1}{2}V$ is a neighbourhood of 0 in (E,C,u). This shows that (E,C,u) is order-quasibarrelled.

The following example shows that a lattice ideal of an order-quasibarrelled vector lattice need not be of the same sort.

Example 3. Let E be the Banach lattice of all continuous real-valued functions on [0,1] equipped with the supremum norm $\|.\|$ and ordered by the positive cone:

$$C = \{f \in E;\ f(x) \geq 0 \text{ for all } x \in [0,1]\}$$

Let F be the subspace of E consisting of all elements $f \in E$ which vanish in a neighbourhood (depending on f) of x = 0. It is easy to see that F is a lattice ideal in E. We show that F is not order-quasibarrelled under the relative topology on F induced by the norm $\|.\|$. Let

$$V = \{f \in F;\ |f(\tfrac{1}{n})| \leq \tfrac{1}{n} \text{ for all } n \geq 1\}.$$

We show that V is a solid barrel in F, but not a neighbourhood of 0 in F. Clearly V is norm-closed and convex. For any $0 \neq f \in F$, there exists a number a_f with $0 < a_f < 1$ such that $f(x) = 0$ for all $x \in [0,a_f]$. If we choose $\lambda = a_f\|f\|^{-1}$, then $\lambda f \in V$ and so V is absorbing in F. It is easy to check that V is solid. Thus V is a solid barrel in F. We now show that V is not a neighbourhood of 0 in F. Let $N \geq 1$ be any natural

number and consider two closed disjoint subsets $[0,\frac{1}{4N}]$ and $[\frac{1}{N+1},1]$ of $[0,1]$. By Urysohn's Lemma, there exists a continuous real-valued function f_N on $[0,1]$ with range in $[0,\frac{1}{N}]$ such that
$$f_N(x) = \begin{cases} 0 & \text{for } x \in [0,\frac{1}{4N}] \\ \frac{1}{N} & \text{for } x \in [\frac{1}{N+1},1] \end{cases}$$
. Clearly $f_N \in F$ and $\|f_N\| = \frac{1}{N}$. On the other hand, since

$$f_N(\frac{1}{N+1}) = \frac{1}{N} > \frac{1}{N+1},$$

it follows that $f_N \notin V$. It then follows that 0 is not an interior point of V and hence V is not a neighbourhood of 0 in $(F,\|.\|)$. This shows that F is not order-quasibarrelled.

Remark 2. F is also an example of a quasibarrelled l.c. vector lattice which is not order-quasibarrelled.

The following proposition shows that under certain conditions, the hereditary property is satisfied for order-quasibarrelled vector lattices.

A lattice ideal I in a vector lattice (E,C) is a σ-normal subspace if it follows from $\sup_n x_n = x \in E$ with $0 \leq x_n \in I$ for all $n \geq 1$ that $x \in I$.

Proposition 10. Let (E,C,u) be a σ-Dedekind complete order-quasibarrelled vector lattice and F a σ-normal subspace of E. Then F is order-quasibarrelled under the induced topology.

Proof: Let V be a solid barrel in F and let

$$U = \{x \in E;\ y \in V \text{ whenever } 0 \leq y \leq |x| \text{ and } y \in F\}.$$

Then U is a closed, convex and solid set in E such that $U \cap F = V$. Furthermore, U is absorbing because otherwise there exists an element $x \varepsilon C$ which is not absorbed by U. Hence, for each positive integer n, there exists $y_n \varepsilon F$ such that $0 \leq y_n \leq \frac{1}{n} x$ and $y_n \notin V$. Since E is σ-Dedekind-complete, $y = \sup\{ny_n; n \geq 1\}$ exists in E. Since F is a σ-normal subspace of E, $y \varepsilon F$. Thus $\{n y\}_{n=1}^{\infty}$ is contained in the order-interval [0,y] in F and is not absorbed by V. But this is absurd, because V is a solid barrel in F. Thus U absorbs every element of E and hence is a solid barrel in E. U must then be a neighbourhood of 0 in (E,C,u). But $V = U \cap F$ and so it follows that V is a neighbourhood of 0 in F. This proves that F is order-quasi-barrelled.

<u>Proposition 11</u>. Let (E,C,u) be an order-quasibarrelled vector lattice and (F,K,v) any l.c. vector lattice. Let f be a continuous almost open (or in particular open) positive linear map of E into F. Then (F,K,v) is order-quasibarrelled.

<u>Proof</u>: Let V be any order-bornivorous barrel in (F,K,v). Then $f^{-1}(V)$ is a barrel in (E,C,u). Furthermore, since f is positive, it follows that $f^{-1}(V)$ is order-bornivorous. Since (E,C,u) is order-quasibarrelled, it follows that $f^{-1}(V)$ is a neighbourhood of 0 in (E,C,u). Since f is almost open, $\overline{f(f^{-1}(V))}$ is a neighbourhood of 0 in F. Hence, it follows from $\overline{f(f^{-1}(V))} \subset \overline{V} = V$ that V is a neighbourhood of 0 in (F,K,v). This proves that (F,K,v) is order-quasibarrelled.

As a particular case, we have the following:

Corollary 9. Let (E,C,u) be an order-quasibarrelled vector lattice and F a closed lattice ideal in E. Then E/F is order-quasibarrelled.

Proposition 12. Let (E,C,u) and (F,K,v) be l.c. vector lattices and f a lattice homomorphism of E into F.

(a) If (E,C,u) is order-quasibarrelled, then f is almost continuous.

(b) If (F,K,v) is order-quasibarrelled and f onto, then f is almost open.

Proof: (a) Let V be a closed, convex and solid neighbourhood of 0 in F. Then $\overline{f^{-1}(V)}$ is an order-bornivorous barrel in (E,C,u) and hence a neighbourhood of 0. This shows that f is almost continuous.

(b) Let W be a closed, convex and solid neighbourhood of 0 in E. Then $\overline{f(W)}$ is a solid barrel in (F,K,v) and hence a neighbourhood of 0. This proves that f is almost open.

§3. Closed graph and Banach Steinhaus Theorems.

Definition 3. An ordered l.c. space (E,C,u) is called an order-quasibarrelled vector space if each order-bornivorous barrel in (E,C,u) is a neighbourhood of 0.

Theorem 4. Let (E,C,u) be an ordered l.c. space for which

$E' \subseteq E^b$ and C is generating. Then the following statements are equivalent:

(a) (E,C,u) is order-quasibarrelled.

(b) Each $O(E',E)$-bounded subset of E' is equicontinuous.

Proof: This is clear if we observe that a subset V of E is order-bornivorous iff V absorbs all order intervals of the form $[-t,t]$, $t \in C$, and this is the case iff V° is an $O(E',E)$-bounded subset of E'.

Theorem 5. Let (E,C,u) be an ordered l.c. space for which order-bounded subsets are u-bounded. Then the following statements are equivalent:

(a) (E,C,u) is order-quasibarrelled.

(b) For any Fréchet space (F,v), a linear map f of E into F is continuous if it satisfies the following two conditions:

(i) f sends each order-bounded subset of E into an u-bounded subset of F;

(ii) the graph of f is closed.

Proof: (a)⇒(b). Let V be a circled, convex neighbourhood of 0 in (F,v). Then $\overline{f^{-1}(V)}$ is an order-bornivorous barrel in E and hence a neighbourhood of 0 in (E,C,u). This shows that f is almost continuous and hence (Chapter 1, Theorem 7(a)) applies to give the continuity of f.

(b)⇒(a). Let W be an order-bornivorous barrel in E, and let p be the gauge of W. Then p is lower semi-continuous. If $(F,\hat{p})$ denotes the completion of the normed vector space $(E/p^{-1}(0),p)$, then $(F,\hat{p})$ is a Banach space and a fortiori a Fréchet space. If g denotes the canonical map of E onto $E/p^{-1}(0)$, then g sends order-bounded subsets of E into $\hat{p}$-bounded subsets of F, because order-bounded subsets of E are u-bounded. We now show that the graph of g is closed. Let $\{x_\alpha\}$ be a net in E converging to $x \in E$ and let $g(x_\alpha)$ converge to y in $(F,\hat{p})$. Then for any $\varepsilon > 0$, there exists $x_1 \in E$ such that

$$\hat{p}(y - g(x_1)) \leq \frac{\varepsilon}{2}.$$

Since p is lower semi-continuous with respect to u and since $\hat{p}$ is continuous with respect to $\hat{p}$-topology, it follows that

$$p(x-x_1) \leq \liminf p(x_\alpha - x_1) = \liminf \hat{p}(g(x_\alpha)-g(x_1)) = \hat{p}(y-g(x_1)).$$

Hence it follows from

$$\hat{p}(y-g(x)) \leq \hat{p}(y-g(x_1)) + \hat{p}(g(x_1)-g(x)) \leq \hat{p}(y-g(x_1)) + p(x_1-x) \leq \varepsilon,$$

that $y = g(x)$. Therefore the graph of g is closed. Hence, by assumption, g is continuous. If B denotes the closed unit ball in $(F,\hat{p})$, then $g^{-1}(B) = W + p^{-1}(0) \subseteq W$. Hence W is a neighbourhood of 0 in E, which proves that (E,C,u) is order-quasibarrelled.

<u>Corollary 10</u>. Let (E,C,u) be an ordered l.c. space with normal cone C. Then the statements (a) and (b) of Theorem 5 are equivalent.

<u>Corollary 11</u>. Let (E,C,u) be an l.c. vector lattice. Then the statements (a) and (b) of Theorem 5 are equivalent.

We now obtain an analogue of the Banach-Steinhaus Theorem for order-quasibarrelled vector spaces and use it to obtain certain results about positive bases. All the results that follow are due to the authors [15].

<u>Theorem 6</u>. (a) Let (E,C,u) be an order-quasibarrelled vector space and (F,K,v) any ordered l.c. space with normal cone K. Let H be a pointwise bounded set of continuous positive linear maps of E into F. Then H is equicontinuous.

(b) Let (E,C,u) be an order-quasibarrelled vector lattice and (F,K,v) any l.c. vector lattice. Let H be a pointwise bounded set of continuous lattice homomorphisms of E into F. Then H is equicontinuous.

<u>Proof</u>: Let V be a closed, convex, circled and full neighbourhood of 0 in (F,K,v) and let $W = \bigcap_{f \in H} f^{-1}(V)$. Then W is a barrel in (E,C,u). Furthermore, since H is pointwise bounded, it easily follows that W is order-bornivorous. Hence W is a neighbourhood of 0 in (E,C,u) which proves that H is equicontinuous.

(b) Let V be a closed convex, solid neighbourhood in (F,K,v) and let $W = \bigcap_{f \in H} f^{-1}(V)$. Then clearly W is a solid barrel in

(E,C,u) and hence, by Theorem 1, W is a neighbourhood of 0 which proves that H is equicontinuous.

As a particular case, we have the following:

Corollary 12. (a) Let (E,C,u) and (F,K,v) be as in Theorem 6(a). If $\{f_n; n \geq 1\}$ is a pointwise bounded sequence of continuous positive linear maps of E into F, then it is equicontinuous.
(b) Let (E,C,u) and (F,K,v) be as in Theorem 6(b). If $\{f_n; n \geq 1\}$ is a pointwise bounded sequence of continuous lattice homomorphisms of E into F, then it is equicontinuous.

Corollary 13. (a) Let (E,C,u) and (F,K,v) be as in Theorem 6(a) and let K be closed in F. If $\{f_n; n \geq 1\}$ is a sequence of continuous positive linear maps, converging pointwise to $f: E \to F$, then f is continuous, positive and linear.
(b) Let (E,C,u) and (F,K,v) be as in Theorem 6(b). If $\{f_n; n \geq 1\}$ is a sequence of continuous lattice homomorphisms of E into F, converging pointwise to $f: E \to F$, then f is a continuous homomorphism.

The following theorem is an analogue of the so called weak basis theorem for an order-quasibarrelled vector space with a positive basis.

An u-Schauder basis $\{x_i, f_i\}$ in an ordered l.c. space (E,C,u) is called a positive u-Schauder basis if $\{x_i\} \subset C$ and all f_i's are positive.

Theorem 7. Let (E,C,u) be an order-quasibarrelled vector space with normal cone C. Then every positive $\sigma(E,E')$-Schauder basis in E is an u-Schauder basis.

<u>Proof</u>: Let $\{x_i, f_i\}$ be a positive $\sigma(E,E')$Schauder basis in E. Define the maps T_n: $E \to E$ by $T_n(x) = \sum_{i=1}^{n} f_i(x)x_i$, $x \in E$, $n \geq 1$. Clearly each T_n is continuous, positive and linear. Furthermore $\lim_n T_n(x) = x$ in the weak topology of E; hence the sequence $\{T_n; n \geq 1\}$ is pointwise bounded. Hence $\{T_n; n \geq 1\}$ is equicontinuous by Corollary 12(a). The rest of the proof is the same as that of (Chapter 2, Theorem 5).

<u>Corollary 14</u>. Let (E,C,u) be an order-quasibarrelled vector space with the normal cone C such that C has non-empty interior. Then each positive $\sigma(E,E')$-basis in E is an u-Schauder basis.

<u>Proof</u>: Let $\{x_i, f_i\}$ be a positive $\sigma(E,E')$-basis in E. Since the positive cone C has non-empty interior, it follows that each f_n is in E' (Chapter 1, Corollary 11). Hence $\{x_i, f_i\}$ is a positive $\sigma(E,E')$-Schauder basis in E so that Theorem 7 applies to give the required result.

Now we prove the analogue of the isomorphism theorem (Chapter 2, Theorem 6) for order-quasibarrelled vector spaces with positive Schauder basis.

<u>Theorem 8</u>. Let (E,C,u) and (F,K,v) be order-quasibarrelled vector spaces with the cones C and K both closed and normal in E and F respectively. Let $\{x_i, f_i\}$ and $\{y_i, g_i\}$ be positive Schauder basis in E and F respectively. Then $\{x_i, f_i\}$ is similar to $\{y_i, g_i\}$ iff there exists a positive isomorphism T: $E \to F$ such that $Tx_i = y_i$ for all $i = 1,2,\ldots$.

Proof: If such a T exists, then similarity follows directly. For the converse, assume that the bases are similar. For $x \in E$, $x = \sum_{i=1}^{\infty} f_i(x)x_i$. Define T_n by $T_n(x) = \sum_{i=1}^{n} f_i(x)y_i$, $n \geq 1$ and T by $T(x) = \sum_{i=1}^{\infty} f_i(x)y_i$. Clearly T is well-defined, 1-1, onto and each T_n is continuous, positive, linear and $\{T_n\}$ converges pointwise to T. Hence, by Corollary 13(a), T is continuous, positive and linear. Similarly T^{-1} is continuous and thus T is the desired positive isomorphism.

CHAPTER V

COUNTABLY BARRELLED AND COUNTABLY QUASIBARRELLED SPACES

We recall that an l.c. space E is a (DF)-space if (a) E has a fundamental sequence of bounded sets, and (b) every strongly bounded subset of E', which is the countable union of equicontinuous subsets of E', is itself equicontinuous. This concept of (DF)-space, which is due to Grothendieck [10], led the first author [13] to introduce two new classes of l.c. spaces which he has labelled as the classes of countably barrelled and countably quasibarrelled spaces. They generalize the well known classes of barrelled and quasibarrelled spaces respectively.

§1. Definitions, characterizations, and relationships.

All the results of this section, unless otherwise stated, are due to the first author [13].

Definition 1. An l.c. space E is said to be a countably barrelled space if each $\sigma(E',E)$-bounded subset of E', which is the countable union of equicontinuous subsets of E', is itself equicontinuous.

This definition coupled with the well-known theorem of Alaoglu-Bourbaki (Theorem 9, §7, Chapter I) immediately leads us to the following:

<u>Proposition 1</u>. Let E be a countably barrelled space. Then every subset of E', which is the countable union of equicontinuous subsets of E', is weakly relatively compact iff it is weakly bounded.

<u>Definition 2</u>. Let E be an l.c. space. A barrel (bornivorous barrel) B which is the countable intersection of circled, convex and closed neighbourhoods of 0 in E is called an N-barrel (bornivorous N-barrel).

<u>Theorem 1</u>. An l.c. space E is countably barrelled iff each N-barrel in E is a neighbourhood of 0.

<u>Proof</u>: Suppose that E is countably barrelled. Let $B = \bigcap_{n=1}^{\infty} V_n$ be an N-barrel in E. Then $\bigcup_{n=1}^{\infty} V_n^o \subset B^o$ and hence $\bigcup_{n=1}^{\infty} V_n^o$ is $\sigma(E',E)$-bounded, because B being absorbing implies B^o is $\sigma(E',E)$-bounded. Since each V_n is a neighbourhood of 0 in E, V_n^o is equicontinuous and hence, by Definition 1, $\bigcup_{n=1}^{\infty} V_n^o$ is equicontinuous. Therefore,

$$\left(\bigcup_{n=1}^{\infty} V_n^o\right)^o = \bigcap_{n=1}^{\infty} V_n^{oo} = \bigcap_{n=1}^{\infty} V_n = B$$

is a neighbourhood of 0 in E.

Conversely, suppose that the condition is satisfied. Let $H = \bigcup_{n=1}^{\infty} H_n$ be a $\sigma(E',E)$-bounded subset of E', where each H_n is an equicontinuous subset of E'. Then

$$B = \left(\bigcup_{n=1}^{\infty} H_n\right)^o = \bigcap_{n=1}^{\infty} H_n^o$$

is an N-barrel in E. Hence B is a neighbourhood of 0 in E and so $B^{\circ} = (\bigcup_{n=1}^{\infty} H_n)^{\circ\circ}$ is equicontinuous. But then $\bigcup_{n=1}^{\infty} H_n$ is equicontinuous which proves the theorem.

A consequence of Theorem 1 is the following:

<u>Corollary 1</u>. Every barrelled space is countably barrelled.

Examples of countably barrelled spaces which are not barrelled will be given later on.

<u>Definition 3</u>. An l.c. space E is said to be a countably quasibarrelled space if each $\beta(E',E)$-bounded subset of E', which is the countable union of equicontinuous subsets of E', is itself equicontinuous.

<u>Remark 1</u>. Clearly every (DF)-space is countably quasibarrelled.

<u>Theorem 2</u>. An l.c. space E is countably quasibarrelled iff each bornivorous N-barrel in E is a neighbourhood of 0.

<u>Proof</u>: A subset S of E is bornivorous iff S° is $\beta(E',E)$-bounded, and S is absorbing iff S° is $\sigma(E',E)$-bounded. The proof can now be accomplished as in Theorem 1.

As easy consequences of Theorem 2, we have

<u>Corollary 2</u>. Every quasibarrelled space is countably quasibarrelled.

<u>Corollary 3</u>. Every countably barrelled space is countably quasibarrelled.

Examples will be given later on to show that the

converse in each of the Corollaries 2 and 3 is false.

A sufficient condition for a countably quasibarrelled space to be countably barrelled is given in the following;

Proposition 2. A sequentially complete countably quasi-barrelled space E is countably barrelled.

Proof: Let $B = \bigcap_{n=1}^{\infty} V_n$ be an N-barrel in E. Then, B° is $\sigma(E',E)$-bounded and hence $\beta(E',E)$-bounded, because E is sequentially complete. Hence $B^{\circ\circ} = B$ is bornivorous, and so a neighbourhood of 0 in E which completes the proof.

Corollary 4. A quasicomplete (or complete) countably quasi-barrelled space is countably barrelled.

The following example shows that a countably barrelled (countably quasibarrelled) space need not be barrelled (quasi-barrelled).

Example 1. Let E be any metrizable l.c. space. Then $(E',\beta(E',E))$ is a complete (DF)-space and hence countably barrelled by Remark 1 and Corollary 4. However $(E',\beta(E',E))$ need not be quasibarrelled in view of the example constructed by Köthe ([31],§31,7).

Proposition 3. The completion $\tilde{E}$ of a countably quasibarrelled space E is countably barrelled.

Proof: Let $B = \bigcap_{n=1}^{\infty} V_n$ be an N-barrel in $\tilde{E}$. Clearly, $B \cap E$ is a barrel in E. Let S be a bounded subset of E. Then B absorbs

$\overline{\Gamma}(S)$ where $\Gamma(S)$ is the circled convex hull of S and $\overline{\Gamma}(S)$ is the closure of $\Gamma(S)$ in E. Hence $B \cap E$ absorbs S which implies $B \cap E$ is bornivorous. Furthermore, $B \cap E = \bigcap_{n=1}^{\infty} (V_n \cap E)$, and each $V_n \cap E$ is a neighbourhood of 0 in E. Hence $B \cap E$ is a bornivorous N-barrel in E and so a neighbourhood of 0, by Theorem 2. But then, $\overline{B \cap E} = B$ is a neighbourhood of 0 in E, where the closure of $B \cap E$ is taken in $\tilde{E}$. This proves the proposition.

The following lemma has been proved by Webb [56] for a more general class of spaces (see Chapter 6 for proof). Using this, a sufficient condition can be given for a countably barrelled space to be barrelled.

Lemma 1. Let E be a countably barrelled space. Then every $\sigma(E',E)$-bounded subset of E' is $\beta(E',E)$-bounded.

Proposition 4. A countably barrelled space E is barrelled iff it is quasibarrelled.

Proof: If E is barrelled, obviously it is quasibarrelled. The converse is clear in view of Lemma 1.

Since a metrizable l.c. space is bornological which, in turn, is quasibarrelled, we have the following:

Corollary 5. Every metrizable (or even bornological) countably barrelled space is barrelled.

<u>Example 2</u>. Let ϕ be the normed vector space of all sequences with only finitely many non-zero coordinates, equipped with the supremum norm. Clearly ϕ is quasibarrelled (because it is normed) and hence countably quasibarrelled. But ϕ is not countably barrelled; for if it were, then by Corollary 5, it would be barrelled which is not true (Chapter 2, Example 2).

<u>Remark 2</u>. It is now clear that the results which are true for metrizable barrelled spaces gain nothing in generality if we consider metrizable countably barrelled spaces.

A t.v. space with a basis is separable and we show below that a separable countably barrelled space is barrelled. Hence all those results about bases, which are valid for barrelled spaces gain nothing in generality if we consider countably barrelled spaces. However, the situation will be different for extended bases, and this will be treated in the next chapter for a more general class of spaces.

The following proposition is due to De Wilde and Houet [7] who have proved it for a more general class of spaces (see Chapter 6). All the remaining results of this section are, in fact, due to De Wilde and Houet [7].

<u>Proposition 5</u>. Let E be a countably barrelled (countably quasibarrelled) space and M a separable subspace of E. Then any barrel (respectively bornivorous barrel) in E induces a neighbourhood of 0 in M.

Proof: Let B be a barrel (respectively bornivorous barrel) in E. Then the set $M\backslash B$ is open in M and hence separable. Let $D = \{x_n; n \geq 1\}$ be a countable dense subset of $M\backslash B$. For each n, $x_n \notin B$ and hence there exists $x'_n \in E'$ such that $\langle x_n, x'_n \rangle > 1$ and $x'_n \in B^0$. Let $H = \bigcup_{n=1}^{\infty} \{x'_n\}$. Of course, $B \subset H^0$ and hence $H^0 = \bigcap_{n=1}^{\infty} \{x'_n\}^0$ is an N-barrel (or a bornivorous N-barrel) in E. Since E is countably barrelled (countably quasibarrelled) H^0 is a neighbourhood of 0 in E, and $\mathrm{int}(H^0) \neq \emptyset$. Since $D \subset M\backslash \mathrm{int}(H^0)$, we have

$$M\backslash B \subset M \cap \bar{D} \subset M \setminus \mathrm{int}(H^0).$$

Therefore, $\mathrm{int}(H^0) \cap M \subset B$ and so $B \cap M$ is a neighbourhood of 0 in M. This completes the proof.

Corollary 6. If E is separable and countably barrelled (countably quasibarrelled) then it is barrelled (quasibarrelled).

Corollary 7. Let (E,u) be a metrizable l.c. space such that $(E',\beta(E',E))$ is separable. Then $(E',\beta(E',E))$ is bornological.

Proof: $(E',\beta(E',E))$ is a complete (DF)-space (Chapter 2, Remark before Theorem 16) and hence by Remark 1 and Corollary 4, it is countably barrelled. But then by Corollary 6, $(E',\beta(E',E))$ is barrelled, and so bornological by (Chapter 2, Corollary 17).

Corollary 8. If E is countably barrelled (countably quasi-barrelled) and E_t is E equipped with the topology defined by its barrels (bornivorous barrels), then a sequence converges in E iff it converges in E_t.

Proof: Let $\{x_n; n \geq 1\}$ be a sequence in E and $x \in E$. Then the linear hull L of x and the x_n is a separable subspace of E; hence the topologies induced by E and E_t are equivalent in L. Hence x_n converges to x in E_t iff x_n converges to x in E.

Definition 4. Let E be an l.c. space. A sequence $\{E_n; n \geq 1\}$ of sets in E is called an absorbing sequence if

(i) E_n is circled and convex for each n;

(ii) $E_n \subset E_{n+1}$ for each n and

(iii) $\bigcup_{n=1}^{\infty} E_n$ is absorbing.

It is called a bounded-absorbing sequence if, in addition, each bounded set $B \subset E$ is absorbed by some E_n.

Corollary 9. Let E be a countably barrelled (countably quasi-barrelled) space. If there exists an absorbing (a bounded-absorbing) sequence of metrizable subsets in E, then E is barrelled (quasibarrelled).

Proof: Let $\{E_n; n \geq 1\}$ be an absorbing (a bounded-absorbing) sequence of metrizable sets in E. Then, for each n, nE_n is metrizable. Let B be a barrel (bornivorous barrel) in E; then any convergent sequence of E converges also for the semi-norm gauge of B, by Corollary 8. Hence, since nE_n is metrizable, $B \cap nE_n$ is a neighbourhood of 0 in nE_n. The proof is now complete in view of Proposition 5.

Theorem 3. Let E be a countably barrelled (countably quasibarrelled) space $\{E_n;\ n \geq 1\}$ an absorbing (a bounded-absorbing) sequence in E and $\lambda_n \uparrow \infty$. Then each circled and convex set V such that for each n, $V \cap \lambda_n E_n$ is a neighbourhood of 0 in $\lambda_n E_n$ for the topology induced by E, is a neighbourhood of 0 in E.

<u>Proof</u>: For each n, let V_n be a neighbourhood of 0 in E such that $V_n \cap \lambda_n E_n \subset V$. Let

$$B = \bigcap_{n=1}^{\infty} \overline{(V \cap \lambda_n E_n) + V_n}.$$

Then B is an N-barrel (a bornivorous N-barrel) provided we show that B is absorbing (bornivorous) in E. Let us prove that B is bornivorous. If A is a bounded set in E, then $A \subset \lambda_{n_0} E_{n_0}$ for some n_0 and there exists $\lambda > 0$ such that $A \subset \lambda V_{n_0}$. Hence

$$A \subset \operatorname{Sup}(1,\lambda)\ (V_{n_0} \cap \lambda_{n_0} E_{n_0}) \subset \operatorname{Sup}(1,\lambda)(V \cap \lambda_n E_n)$$

for each $n \geq n_0$. Moreover, there exists μ such that $A \subset \mu V_n$ for each $n < n_0$. Hence

$$A \subset \operatorname{Sup}(1,\lambda,\mu) B.$$

Thus B is an N-barrel (a bornivorous N-barrel) in E and hence a neighbourhood of 0. We conclude the proof by showing that $B \subset 3V$. Let $x \in B$. For some n_0, $x \in \lambda_{n_0} E_{n_0}$. We also have $x \in B \subset (V \cap \lambda_{n_0} E_{n_0}) + 2V_{n_0}$. Hence

$$x = y + z,\ y \in V \cap \lambda_{n_0} E_{n_0},\ z \in 2V_{n_0}.$$

Since $x \in \lambda_{n_0} E_{n_0}$, it follows that

$$z \in 2(V_{n_0} \cap \lambda_{n_0} E_{n_0}) \subset 2V$$

and hence $x \in 3V$.

The following corollary is an immediate consequence of Theorem 3.

Corollary 10. Let (E,u) be a countably barrelled (countably quasibarrelled) space. If E is the union of an increasing sequence (respectively bounded-absorbing sequence) of vector subspaces E_n, it is the inductive limit of the E_n's equipped with the topology induced by E.

§2. C(X), countably barrelled and countably quasibarrelled spaces.

Throughout in this section, the space C(X) of all continuous real-valued functions on a completely regular Hausdorff space X is assumed to have compact-open topology.

We recall that if $\Psi \in C(X)'$, then the support of Ψ, written $\text{supp}(\Psi)$, is the smallest closed subset A of X such that $\Psi(f) = 0$ for every $f \in C(X)$ vanishing on A; $\text{supp}(\Psi)$ always exists and is compact. If $B \subseteq C(X)'$, supp(B) is defined by

$$\text{supp}(B) = \text{cl}(\bigcup\{\text{supp}(\Psi);\ \Psi \in B\}).$$

Most of the results in this section are due to Morris and Wulbert [37].

Lemma 2. Let X be a completely regular Hausdorff space. A subset B of C(X)' is equicontinuous iff the support K of B is compact and

$$\sup\{|\Psi(f)| : \Psi \in B,\ f \in C(X),\ p_K(f) \leq 1\} < \infty.$$

Proof: Suppose that B is equicontinuous. Then B is contained in the polar of some neighbourhood of 0 in C(X). That is, for some $n > 0$ and some compact set $H \subset X$,

$B \subset n\{\Psi \in C(X)' : \ |\Psi(f)| \leq 1$ for every $f \in C(X)$ with $p_H(f) \leq 1\}$.

Let $\Psi \in B$ and suppose that $f \in C(X)$ and f vanishes on H. Then $p_H(mf) = 0 \leq 1$, for all $m > 0$. Hence $m|\Psi(f)| \leq 1$ for all $m > 0$ implies $\Psi(f) = 0$. We conclude that $\mathrm{supp}(\Psi) \subset H$ and hence $\mathrm{supp}(B) \subset H$ and is therefore compact. Let $K = \mathrm{supp}(B)$ and let $f \in C(X)$ be such that $p_K(f) = 1$. There exists an extension $\tilde{f} \in C(X)$ of the restriction of f to K such that $p_H(\tilde{f}) \leq 1$. Then $|\Psi(\tilde{f})| \leq n$, for all $\Psi \in B$. On the other hand, $f-\tilde{f}$ vanishes on K so that $\Psi(f) = \Psi(\tilde{f})$ for all $\Psi \in B$. Hence $\sup\{|\Psi(f)| : \Psi \in B,\ f \in C(X),\ p_K(f) \leq 1\} \leq n$. Conversely, let

$$r = \sup\{|\Psi(f)| : \Psi \in B,\ f \in C(X),\ p_K(f) \leq 1\}.$$

Clearly, $B \subset r\{\Psi \in C(X)' : |\Psi(f)| \leq 1$ for all $f \in C(X)$ with $p_K(f) \leq 1\}$. Hence B is contained in the polar of a neighbourhood of 0 in C(X) and is therefore equicontinuous. This completes the proof.

Definition 5. A subset A of a topological space X is said to be C(X)-pseudocompact if every function in C(X) is bounded on A.

Remark 3. It is clear that every pseudocompact space is C(X)-

-pseudo compact. But a C(X)-pseudocompact set need not be a pseudocompact space.

The proof of the following proposition, which is <u>omitted</u>, is a slight variation of that of a result of Nachbin [38].

<u>Proposition 6</u>. Let X be a completely regular Hausdorff space. The support of every weakly bounded (hence weakly compact) subset of C(X)' is C(X)-pseudocompact. Further, C(X) is barrelled iff the support of each such set is compact.

Now we prove a theorem which helps us construct another example of a countably barrelled space which is not barrelled.

<u>Theorem 4</u>. Let X be a completely regular Hausdorff space. C(X) is countably barrelled iff every C(X)-pseudocompact subset of X which is the closure of a countable union of compact sets is actually compact.

<u>Proof</u>: Suppose that C(X) is countably barrelled. Let B be a C(X)-pseudocompact subset of X such that $B = cl(\bigcup_{n=1}^{\infty} K_n)$, where each K_n is a compact subset of X. For each $n \geq 1$, let $A_n = \{\Psi \in C(X)';\ supp(\Psi) \subset K_n$ and $|\Psi(f)| \leq 1$ for every $f \in C(X)$ with $p_{K_n}(f) \leq 1\}$.

By Lemma 2, each A_n is equicontinuous. Let $A = \bigcup_{n=1}^{\infty} A_n$. Clearly A is $\sigma(C(X)',C(X))$-bounded and so supp(A) = B is compact by Lemma 2. Conversely, suppose that the condition is satisfied. If C(X) is not countably barrelled, then there exists a bounded set $A = \bigcup_{n=1}^{\infty} A_n$, where each A_n is an equicontinuous subset of $C(X)'$ but A is not equicontinuous. Clearly, supp(A) = $cl(\bigcup_{n=1}^{\infty} supp(A_n))$. By Lemma 2, $supp(A_n)$ is compact and supp(A) is not compact, although supp(A) is C(X)-pseudocompact by Proposition 6. Thus we have a contradiction to our assumption. Hence C(X) must be countably barrelled.

Example 3. Let W be the space of ordinals less than the first uncountable ordinal. Since W is pseudocompact and not compact, C(X) is not barrelled (Proposition 6). However, C(W) is countably barrelled by Theorem 4, because the closure of a countable union of compact subsets of W is compact.

Not every C(X) is countably barrelled as is clear from the following:

Example 4. Let W* be the space of ordinals less than or equal to the first uncountable ordinal, and let T be the Tychonov plank. Then

$$T = cl(\bigcup_{n=1}^{\infty} (W^* \times \{n\})) \subset W^{*2}$$

Clearly each $W^* \times \{n\}$ is compact and T is pseudocompact, but not compact. Hence C(T) is not countably barrelled.

The following theorem, whose proof is omitted, is due to Warner ([55], Theorem 8).

Theorem 5. Let X be a completely regular Hausdorff space. The following statements are equivalent:

(i) C(X) is quasibarrelled.

(ii) Every $\beta(C(X)',C(X))$-bounded subset of $C(X)'$ contained in $\hat{X}$ is equicontinuous.

(iii) For every closed non-compact subset $\hat{S}$ of X, there exists a positive real lower semi-continuous function f on X which is bounded on every compact subset of X but unbounded on S.

Remark 4. For the definition of $\hat{X}$, see the proof of Proposition 8 in Chapter 4.

Lemma 3. If B is bounded in $(C(X)',\beta(C(X)',C(X))$, each positive real lower semi-continuous function on X, bounded on each compact subset of X, is bounded on supp(B).

Proof: This follows from the proof of (3)$\Rightarrow$(1) in ([55], Theorem 8).

Theorem 6. The space C(X), equipped with compact-open topology, is countably quasibarrelled iff, for each countable union of compact sets, the closure F(of the union) being non-compact, there exists a positive real lower semi-continuous function on X, bounded on each compact set and unbounded on F.

Proof: To prove the "only if" part, let $K_m \subset X$, $m \geq 1$ be countably many compact sets the closure F of whose union is not compact. The sets $B_m = \{J_x; x \in K_m\}$, $m \geq 1$, where $J_x \in C(X)'$ defined by $J_x(f) = f(x)$, are clearly equicontinuous but, since F is not compact, their union is not equicontinuous (cf: Lemma 2). Since $C(X)$ is countably quasibarrelled, $\bigcup_{m=1}^{\infty} B_m$ is not bounded in $(C(X)', \beta(C(X)', C(X)))$ and there exists a bounded set $B \subset C(X)$ such that

$$(*) \qquad \sup_{f \varepsilon B} \sup_{x \varepsilon \bigcup_{m=1}^{\infty} K_m} |f(x)| = \sup_{f \varepsilon B} \sup_{J \varepsilon \bigcup_{m=1}^{\infty} B_m} |J(f)| = \infty.$$

We then conclude that the function f defined on X by $f(x) = \sup_{g \varepsilon B} |g(x)|$, for each $x \varepsilon X$ is positive real, lower semicontinuous on X and bounded on each compact subset whereas (*) shows that it is not bounded on F because $F \supset \bigcup_{m=1}^{\infty} K_m$.

To prove the "if" part, let $\{B_m\}$ be a countable family of equicontinuous sets such that $\bigcup_{m=1}^{\infty} B_m$ is $\beta(C(X)', C(X))$-bounded. The sets $supp(B_m)$ are then compact. If the closure F of their union is not compact, there exists a positive real lower semicontinuous function on X, bounded on each compact set and unbounded on F, which is a contradiction by Lemma 3. Hence F is compact. But then, $\bigcup_{m=1}^{\infty} B_m$ is $\sigma(C(X)', C(X))$-bounded and $F = supp(\bigcup_{m=1}^{\infty} B_m)$ is compact; hence $\bigcup_{m=1}^{\infty} B_m$ is equicontinuous by Lemma 2. This completes the proof.

We have already shown that every (DF)-space is countably quasibarrelled. Warner ([55], Theorem 12) has shown that $C(X)$ is a (DF)-space iff the closure of the union of any countable family of compact subsets of X is compact.

§3. The Banach-Steinhaus theorem.

The well-known Banach-Steinhaus theorem is already known to be true for barrelled spaces. Here we prove it for countably barrelled spaces. All the results of this section are due to the first author [13].

Proposition 7. Let E be a countably barrelled (or countably quasibarrelled) space and F any l.c. space. Let $\{H_n; n \geq 1\}$ be a sequence of equicontinuous sets of linear maps of E into F such that $H = \bigcup_{n=1}^{\infty} H_n$ is simply (or strongly) bounded. Then H is equicontinuous.

Proof: We establish the result for countably barrelled spaces. With appropriate changes, the result follows for countably quasibarrelled spaces as well. Let V be a circled, convex and closed neighbourhood of 0 in F, and let

$$W_n = \bigcap_{f \in H_n} f^{-1}(V).$$

Then each W_n is a circled, convex and closed neighbourhood of 0 in E. Now, let

$$U = \bigcap_{n=1}^{\infty} W_n = \bigcap_{f \in H} f^{-1}(V).$$

Since H is simply bounded, U is absorbing and hence an N-barrel in E. Hence U is a neighbourhood of 0 in E by Theorem 1. We now conclude that H is equicontinuous because $U = \bigcap_{f \in H} f^{-1}(V)$.

Corollary 11. Let E be a countably barrelled (or countably quasibarrelled) space and F any l.c. space. Let $\{f_n; n \geq 1\}$

be a simply (or strongly) bounded sequence of continuous linear maps of E into F. Then $\{f_n; n \geq 1\}$ is equicontinuous.

Proof: This follows at once from Proposition 7 by taking $H_n = \{f_n\}$.

Corollary 12. Let E be a countably barrelled space and F any l.c. space. Let $\{f_n; n \geq 1\}$ be a sequence of continuous linear maps of E into F such that $\{f_n; n \geq 1\}$ converges simply to a map f_0 of E into F. Then f_0 is a continuous linear map.

Proof: Clearly f_0 is a linear map. Moreover, $\{f_n; n \geq 1\}$ is equicontinuous by Corollary 11. It now follows from (Chapter 1, Corollary 6) that f_0 is continuous.

Theorem 7. (Banach-Steinhaus). Let E be a countably barrelled space and $\{f_n: n \geq 1\}$ a sequence in E'. If $\{f_n: n \geq 1\}$ is $\sigma(E',E)$-convergent to a linear functional f_0, then $f_0 \in E'$ and $\{f_n: n \geq 1\}$ converges to f_0 uniformly on each precompact subset of E.

Proof: That $f_0 \in E'$ follows from Corollary 12, and that $\{f_n: n \geq 1\}$ converges to f_0 uniformly on each precompact subset of E follows from (Chapter 1, Corollary 6).

An important property of countably barrelled spaces, which is established in the following proposition, helps us show that a metrizable countably barrelled space is barrelled.

Proposition 8. Let E be a countably (or σ-) barrelled space. Then

$(E',\sigma(E',E))$ is sequentially complete.

Proof: Let $\{f_n: n \geq 1\}$ be a Cauchy sequence in $(E',\sigma(E',E))$. Then $\{f_n: n \geq 1\}$ is $\sigma(E',E)$-bounded and hence equicontinuous by Corollary 11. Now define $f_0(x) = \lim_n f_n(x)$ for each $x \in E$. Then $f_0 \in E'$ by Corollary 12, and indeed $\{f_n: n \geq 1\}$ converges to f_0 in the topology $\sigma(E',E)$.

Corollary 13. Let E be a countably barrelled space. Then E', endowed with the topology p of uniform convergence on precompact subsets of E, is sequentially complete.

Proof: Since a p-Cauchy sequence $\{f_n: n \geq 1\}$ in E' is $\sigma(E',E)$ Cauchy, it converges to $f_0 \in E'$ weakly. Thus the corollary follows from Proposition 8 and Theorem 7.

An immediate consequence of Corollary 13 is:

Corollary 14. If E is a countably barrelled space with a countable fundamental system of precompact subsets, then (E',p) is a Fréchet space.

§4. Permanence Properties.

Proposition 9. Let E be a countably barrelled space and F any l.c. space. Let f be a linear, continuous and almost open (a fortiori, open) map of E into F. Then F is countably barrelled.

Proof: Let $B = \bigcap_{n=1}^{\infty} V_n$ be an N-barrel in F. Clearly

$$f^{-1}(B) = \bigcap_{n=1}^{\infty} f^{-1}(V_n).$$

Moreover, $f^{-1}(B)$ is an N-barrel in E and hence a neighbourhood of 0 in E. Since f is almost open,

$$\overline{f(f^{-1}(B))} = \overline{B} = B$$

is a neighbourhood of 0 in F, and the proof is complete.

Corollary 15. Let E be a countably barrelled space and M a closed subspace of E. Then the quotient E/M is also countably barrelled.

Theorem 9. Let $\{E_\alpha : \alpha \in A\}$ be a family of countably barrelled spaces, and $(f_\alpha : \alpha \in A)$ a family of linear maps of E_α into a vector space E. Let u be the finest locally convex topology on E for which each f_α is continuous. Then (E,u) is countably barrelled.

Proof: Let $B = \bigcap_{n=1}^{\infty} V_n$ be an N-barrel in E. Clearly

$$f_\alpha^{-1}(B) = \bigcap_{n=1}^{\infty} f_\alpha^{-1}(V_n).$$

Moreover, $f_\alpha^{-1}(B)$ is an N-barrel in E_α and hence a neighbourhood of 0 in E_α for each α. We now conclude that B is a neighbourhood of 0 in E which completes the proof.

As a direct consequence of Theorem 9, we have

Corollary 16. Let $\{E_\alpha : \alpha \in A\}$ be a family of countably barrelled spaces, and E its inductive limit. Then E is also countably barrelled.

Corollary 17. The locally convex direct sum of a family of countably barrelled spaces is again countably barrelled.

Proposition 10. Any product of countably barrelled spaces is again countably barrelled.

Proof: Let $\{(E_\alpha, u_\alpha);\ \alpha \in A\}$ be a family of countably barrelled spaces and B an N-barrel in $(E,u) = \prod_{\alpha \in A}(E_\alpha, u_\alpha)$. If v_α is the finest l.c. topology on E_α then (E_α, v_α) is barrelled and hence $(E,v) = \prod_{\alpha \in A}(E, v_\alpha)$ is barrelled. Since $v \supseteq u$, B is also an N-barrel in (E,v) and hence a neighbourhood of 0 in (E,v) because an N-barrel is always a barrel. Thus, for some finite subset A_0 of A

(*) $$\prod_{\alpha \in A \setminus A_0} E_\alpha \subset B.$$

Clearly $\prod_{\alpha \in A_0} E_\alpha$ is countably barrelled. Hence $B \cap \prod_{\alpha \in A_0} E_\alpha$ is a neighbourhood of 0 in $\prod_{\alpha \in A_0} E_\alpha$. It follows easily from this and (*) that B is a neighbourhood of 0 in (E,u). This completes the proof.

Remark 5. All the results proved so far in this section remain valid for countably quasibarrelled spaces as well.

The following example, which is due to Iyahen [18], demonstrates that the property of being countably barrelled (countably quasibarrelled) does not pass on to closed subspaces. Since any Hausdorff l.c. space is a closed subspace of some barrelled space, it is enough to give an example of a Hausdorff l.c. space which is not countably quasibarrelled (see Prop.8, Ch.2).

Example 5. Let (E,u) be the space c_0 of all null-sequences with the supremum norm topology u, and let v be the associated weak topology on c_0. For each n, let g_n be the linear map of (E,v) into (E,u) defined by $g_n(x) = (x_1,x_2,\ldots,x_n,0,0,\ldots)$. Then $\{g_n: n \geq 1\}$ is a sequence of continuous linear maps of (E,v) into (E,u) such that, for each $x \in E$, $g_n(x) \to x$ in (E,u). Moreover, $\{g_n: n \geq 1\}$ is uniformly bounded on bounded sets. For, if B is the unit ball in (E,u), the union of $g_n(B)$ is contained in B. But $\{g_n: n \geq 1\}$ is not equicontinuous because v is strictly coarser than u. Hence (E,v) is not countably quasibarrelled.

The following results of this section are due to Webb [58].

Lemma 4. Let E be a countably barrelled (countably quasibarrelled) space, and M a closed subspace of E of countable codimension (respectively of countable codimension and such that for each bounded subset B of E, M is of finite codimension in the span of $\{M \cup B\}$). Then M is countably barrelled (respectively countably quasibarrelled).

Proof: Let $\{x_n\}$ be a sequence in E forming a basis for a complementary subspace G of M. Put $E_1 = M$ and $E_n = \text{span}\{E_{n-1},x_{n-1}\}$, $n > 1$ so that $E = \bigcup_{n=1}^{\infty} E_n$. Clearly each E_n is closed because M is closed. Let $P: E \to M$ be the projection map parallel to G. Since M is closed and of finite codimension in E_n, the restriction map $P_n: E_n \to M$ is continuous. Since, by Corollary 10, E is the inductive limit of the sequence $\{E_n\}$, P is continuous. It follows that M has a closed complement in E and that M is isomorphic to a quotient of E by a closed subspace. Hence, by Corollary 15 (Remark 5), M is countably barrelled (countably quasibarrelled).

An immediate consequence of Lemma 4 is the following:

Corollary 17. A closed subspace of finite codimension of a countably quasibarrelled space is countably quasibarrelled.

Proposition 11. If M is a subspace of countable codimension (resp. see Lemma 4) of a countably barrelled (respectively countably quasibarrelled) space, then M is countably barrelled (respectively countably quasibarrelled).

Proof: Since M is dense in $\overline{M}$ and $\overline{M}$ is countably barrelled by Lemma 4, it is sufficient to consider the case where M is dense in E. Let $V = \bigcap_{n\geq 1} V_n$ be an N-barrel in M. Then $\overline{V} \subset \bigcap_{n\,1} \overline{V}_n = W$ (say) and each $\overline{V}_n$ is a neighbourhood of O in E because M is dense in E. Since $(E', \sigma(E',E))$ is sequentially complete, span$(\overline{V})$ is closed (see Lemma 1, Ch.VI) and hence V is absorbing. So W is an N-barrel and therefore a neighbourhood of O in E. Hence $W \cap M = V$ is a neighbourhood of O in M. This completes the proof.

Corollary 18. Let E be a sequentially complete (DF)-space. If M is a countable codimensional subspace of E, then M is a (DF)-space.

Proof: By Remark 1 and Prop. 2, E is countably quasibarrelled. Hence M is countably quasibarrelled by Proposition 11. Clearly M contains a fundamental sequence of bounded sets. Hence M is a (DF)-space.

Let E^+ denote the set of all sequentially continuous linear functionals on E. We know that elements of E' are given by closed hyperplanes in E whereas the elements of E^+ are given by sequentially closed hyperplanes in E.

Lemma 5. Let E be an l.c. space with $E' = E^+$. Let M be a sequentially closed subspace of E such that for every bounded subset B of E, M is of finite codimension in the span, $sp\{M \cup B\}$, of $\{M \cup B\}$. Then M is closed.

Proof: Let $\{x_\alpha;\ \alpha \in A\}$ be a set of points in E linearly independent modulo M, which together with M spans E. For each $\alpha \in A$, define $H_\alpha = M + sp\{x_\beta;\ \beta \in A,\ \beta \neq \alpha\}$. Then H_α is a hyperplane in E. Let $\{a_n\}$ be a sequence in H_α converging to a_0. Then there are points $x_{\beta_1}, \ldots, x_{\beta_m}$ $(\beta_i \in A)$ such that

$$\{a_n\} \subset M + sp\{x_{\beta_1}, \ldots, x_{\beta_m}\} \subset H_\alpha.$$

Since $M + sp\{x_{\beta_1}, \ldots, x_{\beta_m}\}$ is sequentially closed, $a_0 \in H_\alpha$, which shows that H_α is sequentially closed. Since $E' = E^+$, each H_α is closed. But $M = \bigcap_{\alpha\varepsilon A} H_\alpha$ and so M is closed.

Corollary 19. Let E be an l.c. space with $E' = E^+$. If M is a subspace of E such that for every bounded, closed, circled and convex set B, $M \cap B$ is closed, and M is of finite codimension in $sp\{M \cup B\}$, then M is closed.

Proposition 12. Let E be a countably quasibarrelled space with $E' = E^+$. If M is a subspace of E such that $\bar{M}$ is of countable codimension in E, and such that M is of finite codimension in $span\{M \cup B\}$ for each bounded set B, then M is countably quasibarrelled.

Proof: In view of Lemma 4, it is sufficient to consider the case when M is dense in E. Let $V = \bigcap_{n\geq 1} V_n$ be a bornivorous N-barrel in M. Then $\bar{V} \subset \bigcap_{n\geq 1} \bar{V}_n$. Let $G = span\bar{V}$. We first show that $\bar{V}$ is bornivorous in G. Let B be a bounded subset of G. Since $G \supset M$, there exists a finite dimensional subspace H of G such that $B \subset M + H = L \subset G$. Now $\bar{V} \cap L$ is the closure of V in L, M is of finite codimension in L and $\bar{V} \cap L$ is bornivorous in L and so $\bar{V}$ absorbs B. We now show that G = E. Since $G \supset M$ and M is dense in E, it is sufficient to prove that G is closed. Let A be a circled, convex, closed and bounded subset of E. Then for some α, $G \cap B \subset \alpha\bar{V}$ and so $G \cap A$ is closed. Hence G is closed, by Corollary 19. Thus $\bigcap_{n\geq 1} \bar{V}_n$ is a bornivorous N-barrel in E and hence a neighbourhood of 0.

Therefore $V = (\bigcap_{n\geq 1} \overline{V}_n) \cap M$ is a neighbourhood of 0 in M. This proves that M is countably quasibarrelled.

<u>Corollary 20</u>. Let E be a countably quasibarrelled space, $E' = E^+$, and M a subspace of finite codimension in E. Then M is countably quasibarrelled.

<u>§5. H-spaces</u>.

The concept of countably barrelled space permits us to introduce a class of locally convex spaces which generalizes that of distinguished space . All the results of this section are due to the second author [26].

<u>Definition 6</u>. An l.c. space E is said to be an H-space if each $\sigma(E",E')$-bounded subset of its strong bidual E", which is the countable union of equicontinuous subsets of E", is contained in the $\sigma(E",E')$-closure of some bounded subset of E.

<u>Theorem 10</u>. An l.c. space E is an H-space iff $(E',\beta(E',E))$ is countably barrelled.

<u>Proof</u>: Suppose that E is an H-space. Let $B = \bigcup_{n=1}^{\infty} H_n$ be a $\sigma(E",E')$-bounded subset of E" where each H_n is equicontinuous. There exists a bounded subset A of E such that $B \subset A^{\circ\circ}$. Since A° is a neighbourhood of 0 in $(E',\beta(E',E))$, $A^{\circ\circ}$ is equicontinuous and hence B is equicontinuous which implies that $(E',\beta(E',E))$ is countably barrelled. For the converse, let $B = \bigcup_{n=1}^{\infty} H_n$ be a

$\sigma(E'',E')$-bounded subset of E'' such that each H_n is equicontinuous. Then B is equicontinuous and so $B \subset V^\circ$ for some neighbourhood V in $(E',\beta(E',E))$. But then there exists a bounded subset A in E such that $A^\circ \subset V$ and hence $B \subset V^\circ \subset A^{\circ\circ}$. This shows that E is an H-space.

Corollary 21. Every distinguished space (Chapter 2, §3, Def. 5) is an H-space.

Proof: Since a barrelled space is countably barrelled, the result follows at once.

It is known that a metrizable l.c. space need not be distinguished (Chapter 2, §3). However, we show, in the following proposition, that it is an H-space.

Proposition 13. Every metrizable l.c. space is an H-space.

Proof: Let E be a metrizable l.c. space. Then $(E',\beta(E',E))$ is a complete (DF)-space (Chapter 2, §4) and hence countably barrelled by Corollary 4 and Remark 1. Hence E is an H-space in view of Theorem 10.

Köthe ([31],§31,7) has constructed a complete metrizable l.c. space which is not distinguished. However, it is an H-space, in view of Proposition 13. We now give an example of an H-space which is not metrizable.

Example 6. Let ϕ be the space of all sequences with only finitely many non-zero components and let u be the normal topology (see [31], §30) on ϕ; then it is the same as the locally convex direct sum topology on ϕ and hence (ϕ,u) is not metrizable ([12], page 73). But (ϕ,u) is a distinguished space and hence an H-space, because $(\phi^\times,\beta(\phi^\times,\phi)) = (\omega,\beta(\omega,\phi))$ is a Fréchet space ([31], page 408).

In the following proposition, a sufficient condition is given for an H-space to be distinguished.

<u>Proposition 14</u>. Let E be an H-space such that $(E',\beta(E',E))$ is separable. Then E is distinguished.

<u>Proof</u>: Since E is an H-space $(E',\beta(E',E))$ is countably barrelled. It now follows from Corollary 6 that $(E',\beta(E',E))$ is barrelled. Hence E is distinguished.

The following example shows that the converse of Proposition 14 need not be true.

<u>Example 7</u>. The space $\ell^1 = \{x = \{x_i\};\ \sum_{i=1}^{\infty} |x_i| < \infty\}$, with the usual norm $||x|| = \sum_{i=1}^{\infty} |x_i|$, is a distinguished space, because it is a normed space. But its strong dual ℓ^∞ is not separable.

<u>Theorem 11</u>. An H-space E is distinguished iff $(E',\beta(E',E))$ is quasibarrelled.

<u>Proof</u>: If E is distinguished, $(E',\beta(E',E))$ is barrelled and hence quasibarrelled. Conversely suppose that $(E',\beta(E',E))$ is quasibarrelled. Since E is an H-space, $(E',\beta(E',E))$ is countably barrelled. It now follows from Proposition 4 that

$(E',\beta(E',E))$ is barrelled and hence is distinguished.

Corollary 22. Let E be an H-space. If $(E',\beta(E',E))$ is metrizable (or even bornological) then E is distinguished.

That the converse of Corollary 22 need not be true is exhibited in the following:

Example 8. Let ω be the space of all sequences equipped with the normal topology η. Then (ω,η) is a complete metrizable l.c. space whereas its strong dual $(\omega^{\times},\beta(\omega^{\times},\omega)) = (\phi,\beta(\phi,\omega))$ is a non-metrizable barrelled space. Hence (ω,η) is a distinguished space with non-metrizable strong dual.

Since metrizable l.c. spaces are bornological as well as H-spaces, it is interesting to exhibit some relation between bornological and H-spaces. We have the following:

Proposition 15. Let E be a bornological space such that $(E',\beta(E',E))$ is countably quasibarrelled. Then E is an H-space.

Proof: Since E is bornological, $(E',\beta(E',E))$ is complete. It now follows from Corollary 4 that $(E',\beta(E',E))$ is countably barrelled and hence E is an H-space.

We do not know if a bornological space is always an H-space. However, in the opposite direction, the following example shows that an H-space need not be quasibarrelled (and so neither barrelled nor bornological).

Example 9. Let F be a non-reflexive Banach space and

$E = (F', \tau(F',F))$, where $\tau(F',F)$ is the Mackey topology on F'. Then E is semi-reflexive but not reflexive. But an l.c. space is reflexive iff it is semi-reflexive and quasibarrelled (Chapter 2, §1). Hence E is not quasibarrelled. However, E, being semi-reflexive, is a distinguished space and hence an H-space.

§6. Open Problems.

Problem 1. If the conclusion of Proposition 7 is true, is E a countably barrelled (countably quasibarrelled) space?

Problem 2*. Is there an l.c. space E such that $(E, \tau(E,E'))$ is countably barrelled but not barrelled?

Problem 3. Is there a bornological space whose strong dual is not countably barrelled?

Problem 4. (Webb [58]). Is the condition " $E' = E^+$ " in Corollary 20 superfluous?

Problem 5. Is it possible to obtain a result analogous to Theorem 15 of Chapter 2 for H-spaces?

* This problem was communicated to us by Dr. I. Tweddle.

CHAPTER VI

σ-BARRELLED AND SEQUENTIALLY BARRELLED SPACES

§1. σ-barrelled spaces.

An important property of countably barrelled spaces is that each weakly bounded sequence in the dual space forms an equicontinuous set. This has been used by De Wilde and Houet [7] to introduce the concept of σ-barrelled space. Levin and Saxon [32] have also, independently, introduced this concept, which they call ω-barrelled space to study the hereditary properties of subspaces of such spaces.

Definition 1. An l.c. space E is said to be σ-barrelled (σ-quasibarrelled) if each $\sigma(E',E)$-bounded (respectively $\beta(E',E)$-bounded) sequence in E' is equicontinuous.

Remark 1. It is clear from the definitions concerned that a σ-barrelled space is σ-quasibarrelled.

De Wilde and Houet [7] call σ-quasibarrelled spaces σ-evaluable.

Proposition 1. Every countably barrelled (countably quasi-barrelled) space E is σ-barrelled (respectively σ-quasibarrelled).

Proof: If $\{f_n : n \geq 1\}$ is a $\sigma(E',E)$-bounded ($\beta(E',E)$-bounded) sequence in E', we can write

$$\{f_n: \ n \geq 1\} = \bigcup_{n=1}^{\infty} \{f_n\}$$

which is certainly equicontinuous, because E is countably barrelled (respectively countably quasibarrelled). This completes the proof.

Remark 2. It is clear from Definition 1 that the Banach-Steinhaus theorem (Chapter 5, Theorem 7) can be extended to σ-barrelled spaces. As will be shown in section 4, Lemma 1 of Chapter 5 remains valid for σ-barrelled spaces as well and consequently Proposition 4 and Corollary 5 of Chapter 5 extend to σ-barrelled spaces. It is also clear from the proof of Proposition 8 in Chapter 5 that it readily extends to σ-barrelled spaces.

The following theorem is due to Husain and Wong [14].

Theorem 1. An l.c. space E is σ-barrelled (σ-quasibarrelled) iff for any absorbing (bounded absorbing) sequence $\{E_n: \ n \geq 1\}$ in E, the sequence $\{f_n: \ n \geq 1\}$, $f_n \in E^\circ_n$ for all $n \geq 1$, is equicontinuous.

Proof: Suppose that E is σ-barrelled. For any finite subset S of E, there exist $\lambda > 0$ and $n_0 \geq 1$ such that $S \subset \lambda E_n$ for all $n \geq n_0$ and so the sequence $\{f_n; \ n \geq n_0\}$, $f_n \in E^\circ_n$ is $\sigma(E',E)$-bounded and hence equicontinuous; consequently $\{f_n: \ n \geq 1\}$ is equicontinuous. Conversely, assume that the condition is satisfied. Let $\{g_n: \ n \geq 1\}$ be a $\sigma(E',E)$-bounded sequence in E'. For each $m \geq 1$, we define

$$E_m = \{x \in E: \ |g_n(x)| \leq 1 \text{ for all } n \geq m\}.$$

Then $\{E_n: \ n \geq 1\}$ is an absorbing sequence in E. Since $g_m \in E^\circ_m$ for all $m \geq 1$, we conclude that $\{g_m; \ m \geq 1\}$ is equicontinuous. This shows that E is σ-barrelled. Similarly, we can prove for σ-quasibarrelled space.

The following proposition, which is due to De Wilde and Houet [7], describes a property of σ-barrelled spaces, which turns out to be a sufficient condition for σ-quasibarrelled spaces to be σ-barrelled.

Proposition 2. If E is a σ-barrelled space, any absorbing sequence of closed sets is a bounded-absorbing sequence.

Proof: Suppose that $\{C_n; \ n \geq 1\}$ is an absorbing sequence of closed sets in E which is not bounded-absorbing. Then there exists a bounded set $B \subset E$ such that $B \not\subset nC_n$ for all $n \geq 1$. Let $x_n \in B \backslash nC_n$ for each n. Then there exists $f_n \in C^\circ_n$ such that

(*) $$|f_n(x_n)| > n.$$

On the otherhand, $\{f_n: \ n \geq 1\}$ is equicontinuous by Theorem 1 and so it is strongly bounded; consequently $\{f_n: \ n \geq 1\}$ is absorbed by B° which contradicts (*).

Theorem 2. A σ-quasibarrelled space E is σ-barrelled iff each absorbing sequence of closed sets in E is bounded absorbing.

Proof: In view of Remark 1, the "only if" part is always true. To establish the "if" part we proceed as follows: Let

$\{f_n;\ n \geq 1\}$ be a $\sigma(E',E)$-bounded sequence in E'. For each $k \geq 1$, we define

$$E_k = \{x \in E;\ |f_n(x)| \leq 1 \text{ for all } n \geq k\}.$$

Then $\{E_k;\ k \geq 1\}$ is an absorbing sequence of closed sets in E and hence bounded-absorbing. Since E is σ-quasibarrelled and $f_k \in E^\circ{}_k$ for all $k \geq 1$, we conclude, in view of Theorem 1, that $\{f_k;\ k \geq 1\}$ is equicontinuous.

Let E be an l.c. space, B a circled convex and $\sigma(E,E')$-bounded subset of E, and let $E(B) = \bigcup_{n=1}^{\infty} nB$. Define

$$\|x\|_B = \inf\{\lambda > 0:\ x \in \lambda B\},\ x \in E(B).$$

Clearly $\|\cdot\|_B$ is a norm on E(B).

Definition 2. If $(E(B), \|\ \|_B)$ is complete, then B is said to be infracomplete.

Clearly each circled, convex, $\sigma(E,E')$-bounded, and $\sigma(E,E')$-sequentially complete subset of E is infracomplete.

The following result is the Banach-Mackex Theorem:

Theorem 3. Let (E,u) be an l.c. space. Each infracomplete subset B of E is $\beta(E,E')$-bounded.

Proof: Let T be any $\sigma(E',E)$-bounded subset of E'. The topology on E(B) induced by u is coarser than the norm topology and so the restriction f_B on E(B) of each $f \in E'$ is $\|\cdot\|_B$-continuous. Hence the set $\{f_B:\ f \in T\}$ is $\sigma(E(B)',E(B))$-bounded. By appealing to the principle of uniform boundedness we get

$$\sup\{|f_B(x)|; \ f \in T, x \in B\} < \infty$$

which proves that B is $\beta(E,E')$-bounded.

Corollary 1. Let E be an l.c. space and consider the following statements:

(a) E is σ-barrelled.

(b) E has property (C) i.e. each $\sigma(E',E)$-bounded subset of E is $\sigma(E',E)$-relatively countably compact.

(c) E has property (S) i.e. E' is $\sigma(E',E)$-sequentially complete.

(d) Each $\sigma(E',E)$-bounded and $\sigma(E',E)$-closed subset of E' is $\sigma(E',E)$-sequentially complete.

Then (a)$\Rightarrow$(b)$\Rightarrow$(c)$\Rightarrow$(d). Moreover, if E is σ-quasibarrelled, (d)$\Rightarrow$(a).

Proof: The implications (a)$\Rightarrow$(b)$\Rightarrow$(c)$\Rightarrow$(d) are obvious. If E is σ-quasibarrelled, Theorem 3 implies that (d)$\Rightarrow$(a).

Remark 3. Since a σ-barrelled space has property (S), it is easily seen that Corollaries 13 and 14 of Chapter 5 extend to σ-barrelled spaces.

Another consequence of Theorem 3 is the following:

Corollary 2. Let E be an l.c. space such that each $\sigma(E',E)$-bounded $\sigma(E',E)$-closed set in E' is $\sigma(E',E)$-sequentially complete (in particular, E has property (S)).

(i) If E is quasibarrelled, then it is barrelled.

(ii) If E is countably quasibarrelled, then it is countably barrelled.

Levin and Saxon [32] have shown, in the following example, that a Mackey space with property (S) need not have property (C). Hence we conclude from Corollary 1 that Mackey spaces need not be σ-quasibarrelled.

Example 1. Consider the vector space m of all bounded sequences with the Mackey topology $\tau(m,\ell^1)$. Then it is known that $(\ell^1,\sigma(\ell^1,m))$ is sequentially complete so that $(m,\tau(m,\ell^1))$ has property (S). To show that $(m,\tau(m,\ell^1))$ does not have property (C), let $B = \{e_n: n \geq 1\}$ be the canonical Schauder basis of ℓ^1, where each e_n is the sequence having 1 in the nth coordinate and zeros elsewhere. B is $\sigma(\ell^1,m)$-bounded but has no $\sigma(\ell^1,m)$-accumulation point in ℓ^1. For, suppose $y \in \ell^1$. There exists a sequence $s \in m$ of the form $s = (0,\ldots,0,2,2,\ldots)$ such that $|\langle s,y\rangle| < 1$. Thus

$$\langle s,e_n-y\rangle = \langle s,e_n\rangle - \langle s,y\rangle > 1$$

holds for sufficiently large n such that $\langle s,e_n\rangle = 2$. Then $e_n - y \in \{s\}^\circ$ for at most finitely many $n \in \mathbb{N}$. Therefore y is not a $\sigma(\ell^1,m)$-accumulation point of B.

It is clear from (Chapter 5, Example 5) that a subspace of a σ-barrelled (σ-quasibarrelled) space need not be of the same sort. However, Levin and Saxon [32], have shown that a countably codimensional subspace of a σ-barrelled space is

σ-barrelled. The following two Lemmas lead us to that result at once. For a shorter proof of Lemma 1, see Webb [58].

Lemma 1. Let E be an l.c. space with property (S). Let A be a closed, circled and convex subset of E. If the linear span sp(A) of A has countable codimension in E, then sp(A) is closed in E.

Proof: Let $x_1 \in E \setminus sp(A)$ and let $\{x_k;\ k \geq 1\}$ be a linearly independent sequence in E such that $sp(\{x_k\})$ and sp(A) are algebraic supplements to each other. Clearly $A^{\circ\circ} = A$. Since no non-zero scalar multiple of x_k is in A, it follows that A° is unbounded at x_k for $k \geq 1$. Let $\varepsilon > 0$ be given with $\varepsilon < 1$. Let $A_1 = A$ and $A_n = A_{n-1} + \{tx_{n-1};\ |t| \leq 1\}$ for $n \geq 2$. Then $A_n^{\circ\circ} = A_n$ for $n \geq 1$. Choose a sequence $\{f_n\}$ such that $f_n \in (2^{-n}\varepsilon)A_n^\circ$, $f_1(x_1) = 2$, and $\sum_{k=1}^{n+1} f_k(x_{n+1}) = 0$ for $n \geq 1$. This is possible, since A_n° is circled and $\{f(x_n);\ f \in A_n^\circ\}$ is unbounded for $n \geq 1$. Let $x \in E$. Since A_n is absorbing in $sp(A_n)$ and since $\bigcup_{n=1}^\infty sp(A_n) = E$, there is some integer p and a number $\delta > 0$ such that $\delta x \in A_p \subset A_{p+1} \subset A_{p+2} \ldots$. Thus

$$\begin{aligned}\sum_{k=1}^{\infty} |f_k(x)| &= \frac{1}{\delta}\sum_{k=1}^{\infty} |f_k(\delta x)| \\ &= \frac{1}{\delta}\Big(\sum_{k=1}^{p-1} |f_k(\delta x)| + \sum_{k=p}^{\infty} |f_k(\delta x)|\Big) \\ &\leq \frac{1}{\delta}\Big(\sum_{k=1}^{p-1} |f_k(\delta x)| + \sum_{k=p}^{\infty} 2^{-k}\varepsilon\Big) \\ &< \infty.\end{aligned}$$

Since E' is $\sigma(E',E)$-sequentially complete, the linear functional f_ε defined by

$$f_\varepsilon(x) = \sum_{k=1}^{\infty} f_k(x)$$

is continuous on E. Furthermore, for $x \in A_1$,

$$|f_\varepsilon(x)| \leq \sum_{k=1}^{\infty} |f_k(x)| \leq \sum_{k=1}^{\infty} 2^{-k} \cdot \varepsilon = \varepsilon$$

and

$$|f_\varepsilon(x_{n+1})| \leq |\sum_{k=1}^{n+1} f_k(x_{n+1})| + \sum_{k=n+2}^{\infty} |f_k(x_{n+1})|$$

$$\leq 0 + \sum_{k=n+2}^{\infty} 2^{-k} \cdot \varepsilon < \varepsilon$$

for $n \geq 1$. Also, $|f_\varepsilon(x_1)| \geq 2 - \sum_{k=2}^{\infty} |f_k(x_1)| > 2 - \varepsilon > 1$.

Let $g_\varepsilon = (f_\varepsilon(x_1))^{-1} \cdot f_\varepsilon$. Then $g_\varepsilon(x_1) = 1$, $|g_\varepsilon(x)| < \varepsilon$ for $x \in A_1 = A$, and $|g_\varepsilon(x_{n+1})| < \varepsilon$ for $n \geq 1$. Let $\{\varepsilon_n\}$ be a sequence such that $0 < \varepsilon_n < 1$ $(n = 1,2,\ldots)$ and $\varepsilon_n \to 0$. The linear functional h defined on E by

$$h(x) = \begin{cases} 0 & \text{if } x \in A_1, \\ 1 & \text{if } x = x_1, \\ 0 & \text{if } x = x_{n+1}, \text{ for } n \geq 1 \end{cases}$$

is the pointwise limit of a sequence $\{g_{\varepsilon_n};\ n \geq 1\}$ of continuous linear functionals and hence is continuous. Therefore the closure of sp(A) is a subset of $h^{-1}(\{0\})$ and does not contain x_1. It follows that sp(A) is closed.

Corollary 3. Let E be an l.c. space with property (S). Let

M be a dense linear subspace of countable codimension in E. If a subset B of E' is $\sigma(E',M)$-bounded, then B is $\sigma(E',E)$-bounded.

Proof: B° is a closed, circled and convex subset of E. Since B is $\sigma(E',M)$-bounded, $M \subset sp(B°)$. By Lemma 1, sp(B°) is closed. Since M is dense in E, sp(B°) = E. Then B is $\sigma(E',E)$-bounded.

Lemma 2. Let E be an l.c. space with property (S). Let M be a closed vector subspace of countable codimension in E. Then any linear extension to E of a continuous linear functional on M is continuous.

Proof: Let $P = \{x_n ; n \geq 1\}$ be a countable set such that E is the linear span of $M \cup P$. Let M_n be the vector subspace spanned by $M \cup \{x_1,\ldots,x_n\}$. Let f be a continuous linear functional on M and let $\tilde{f}$ be any linear extension of f on E. For each n, M is a closed vector subspace of finite codimension in M_n. Then the restriction f_n of $\tilde{f}$ to M_n is continuous. Let g_n be any continuous extension of f_n to E. An element x of E belongs to M_n for some $n \in \mathbb{N}$. Therefore the sequence $\{g_n(x): n \geq 1\}$ is eventually constant and converges to $\tilde{f}(x)$. Since E has property (S), $\tilde{f}$ is a member of E'. This completes the proof.

Lemma 3. Let M be a dense countable codimensional subspace of a σ-barrelled space E. Then M, equipped with the relative topology, is σ-barrelled.

Proof: M' can be canonically identified with E'. Let $\{f_n; n \geq 1\}$ be a $\sigma(E',M)$-bounded sequence in E'; then it is $\sigma(E',E)$-bounded by Corollary 3. Hence $\{f_n; n \geq 1\}$ is equicontinuous with respect to E and a fortiori, equicontinuous with respect to M. This shows that M is σ-barrelled under the relative topology.

Lemma 4. Let M be a closed countable codimensional subspace of a σ-barrelled space E. Then M, equipped with the relative topology, is σ-barrelled.

Proof: Let p be any algebraic projection of E onto M and $\{f_n; n \geq 1\}$ a $\sigma(M',M)$-bounded sequence in M. In view of Lemma 2, the sequence $\{f_n op; n \geq 1\}$ is in E', and it is easy to see that $\{f_n op: n \geq 1\}$ is equicontinuous in E' which implies that $\{f_n: n \geq 1\}$ is so in M'. Hence M is σ-barrelled under the relative topology.

Theorem 4. Let M be a countable codimensional subspace of a σ-barrelled space E. Then M is σ-barrelled under the relative topology.

Proof: By Lemma 4, $\bar{M}$ is σ-barrelled under the relative topology. But then, by Lemma 3, M is σ-barrelled. This completes the proof.

§2. σ-barrelled spaces and Mackey topology.

This section is devoted to a consideration of conditions

under which a σ-barrelled space is barrelled under its Mackey topology. Most of the results of this section are due to Tweddle [51].

<u>Proposition 3</u>. Let E be a σ-barrelled space. The following statements are equivalent.

(a) E is barrelled under $\tau(E,E')$.

(b) E' is $\tau(E',E)$-quasicomplete.

(c) The $\tau(E,E')$-completion of E is barrelled.

<u>Proof</u>: (a)$\Rightarrow$(b): Since $(E,\tau(E,E')$ is barrelled, E' is σ-(E',E)-quasicomplete and hence also $\tau(E',E)$ quasi-complete.

(b)$\Rightarrow$(c): Let A be any $\sigma(E',E)$-closed, circled, convex and $\sigma(E',E)$-bounded set. Since E is σ-barrelled, each $\sigma(E',E)$-bounded sequence in E' forms a $\sigma(E',E)$-relatively compact set. It now follows from Eberlein's theorem that A is $\sigma(E',E)$-compact so that $\beta(E,E') = \tau(E,E')$.

(c)$\Rightarrow$(a): Let B be any $\sigma(E',E)$-bounded set. If F is the $\tau(E,E')$-completion of E, then B is $\sigma(E',F)$-bounded. If not, there is $x \in F$ and a sequence $\{x'_n\} \subset B$ such that $|\langle x,x'_n\rangle| \geq n$ for all n. Since E is σ-barrelled, $\{x'_n : n \geq 1\}$ is equicontinuous in the initial topology of E and equicontinuous under the finer topology $\tau(E,E')$ and so also under $\tau(F,E')$. Thus, $\sup_n |\langle x,x'_n\rangle| < \infty$ which gives a contradiction. It now follows that B is equicontinuous under $\tau(F,E')$ because F is barrelled. But $\tau(F,E')$ induces $\tau(E,E')$ on E so that B is also $\tau(E,E')$-equicontinuous. This completes the proof.

An immediate consequence of Proposition 3 is the following:

Corollary 4. Let E be a σ-barrelled space. If $(E,\tau(E,E'))$ is quasibarrelled, then $(E,\tau(E,E'))$ is barrelled.

Proposition 4. Let (E,u) be a σ-barrelled space and F the u-completion of E. Suppose that there is a metrizable l.c. topology η on E' which is coarser than $\tau(E',F)$. Then $(E,\tau(E,E')$ is barrelled.

Proof: Let G be the dual of E' under η. Then G is contained in F and $\sigma(E',G)$ is coarser than $\sigma(E',F)$. Let B be any $\sigma(E',E)$-closed, circled, convex and bounded set in E'. As in the proof of Proposition 3, B is $\sigma(E',F)$-bounded and since $\sigma(E',F)$ is finer than $\sigma(E',E)$, B is also $\sigma(E',F)$-closed. The same subsets of E' are equicontinuous under u and the extended topology on F, so that F is also σ-barrelled. It now follows that B is countably compact under $\sigma(E',F)$ and so also under the coarser topology $\sigma(E',G)$. Since $\eta = \tau(E',G)$ is metrizable, B is $\sigma(E',G)$-compact ([31], §24, 3(a); the result is given for a Fréchet space, but it clearly holds for any metrizable space since a weakly countably compact set has the same property in the completion of the space). We now show that $\sigma(E',F)$ and

$\sigma(E',G)$ coincide on B from which it will follow that B is $\sigma(E',F)$-compact and so also $\sigma(E',E)$-compact; this will show that $\beta(E,E') = \tau(E,E')$ which will complete the proof. Let A be any $\sigma(E',F)$-closed subset of B and let x' be an element of the $\sigma(E',G)$-closure of A. Now there is a sequence $(x'_n; n \geq 1) \subset A$ which $\sigma(E',G)$-converges to x' ([31], §24, 1(7)). Since E is σ-barrelled, the $\sigma(E',F)$-closure C of $\{x'_n; n \geq 1\}$ is $\sigma(E',F)$-compact. The topologies $\sigma(E',F)$ and $\sigma(E',G)$, therefore, coincide on C, and this implies that $x' \in C \subset A$ i.e. A is $\sigma(E',G)$-closed. It now follows that the identity map of $(B,\sigma(E',F))$ to $(B,\sigma(E',G))$ is a homeomorphism.

<u>Corollary 5</u>. Let E and F be as in Proposition 4. Suppose that there is a sequence $\{E_n; n \geq 1\}$ of subsets of E such that

(i) $\bigcup_{n=1}^{\infty} E_n$ is total in $(E,\sigma(E,E'))$, and

(ii) for each n, E_n is $\sigma(F,E')$-relatively compact.

Then $(E,\tau(E,E'))$ is barrelled.

<u>Proof</u>: By Krein's theorem ([31],§24, 5(4)), the $\sigma(F,E')$-closed, circled and convex envelope F_n of E_n is $\sigma(F,E')$-compact for each n. Let G be the vector subspace of F generated by $\bigcup_{n=1}^{\infty} F_n$. The topology η on E' of uniform convergence on the sets E_n, $n \geq 1$, is metrizable and under it E' has dual G. Thus $\eta = \tau(E',G)$ which is coarser than $\tau(E',F)$.

<u>Corollary 6</u>. A separable σ-barrelled space is barrelled.

Proof: Let $\{x_n; n \geq 1\}$ be a dense subset of a σ-barrelled space E. If G is the vector subspace of E, generated by this set, the topology $\sigma(E',G)$ on E' is metrizable and clearly satisfies the requirements of Proposition 4. Now a compact metric space is separable and $\sigma(E',G)$ coincides with $\sigma(E',E)$ on each $\sigma(E',E)$-closed bounded set. It follows that each such set, being $\sigma(E',G)$-compact, is $\sigma(E',E)$-separable and therefore equicontinuous. Thus the topology of E is already $\sigma(E,E')$.

Remark 4. De Wilde and Houet [7] have also proved Corollary 6 by a different method which we have adopted in (Chapter 5, Proposition 5, Corollary 6). In fact, as they have shown, Proposition 5 and Corollaries 6 and 8 of Chapter 5 remain valid for σ-barrelled spaces.

Corollary 7. Let E be a metrizable l.c. space. E is reflexive iff E' is σ-barrelled under some topology of the dual pair $\langle E',E\rangle$.

Proof: Let $(V_n; n \geq 1)$ be a countable base of neighbourhoods of 0 in E. Then E' is the union of the polar sets V_n°, $n \geq 1$ which are $\sigma(E',E)$-compact. Thus, if E' is σ-barrelled under some topology of the dual pair $\langle E',E\rangle$, $\beta(E',E) = \tau(E',E)$ by Corollary 5. The "if" part now follows because a metrizable semi-reflexive space is reflexive. The "only if" part is immediate, because if E is reflexive, $(E',\tau(E',E))$ is barrelled and hence σ-barrelled. This completes the proof.

That the initial σ-barrelled topology in Propositions 3 and 4, and Corollary 5 need not be the Mackey topology of the space is clear from the following:

Example 2. Let E be a non-separable Hilbert space. It is countably barrelled and hence σ-barrelled under the topology of uniform convergence on the $\sigma(E',E)$-separable bounded subsets of E'. This is a topology of the dual pair $\langle E, E' \rangle$, but it is clearly strictly coarser than the norm topology under which E is barrelled. In Proposition 4, the required topology η is just $\tau(E',E)$ and in Corollary 5 each E_n may be taken to be the closed unit ball of E.

If a σ-barrelled space E has a countable fundamental system of precompact sets then, by Corollary 5, $(E,\tau(E,E'))$ is barrelled. However, much more can be established directly as has been shown in the following:

Proposition 5. Let E be a σ-barrelled space with a countable fundamental system of precompact sets. Then

(i) E is barrelled.

(ii) Each bounded subset of E is precompact.

(iii) The completion, quasicompletion and bidual of E all coincide.

(iv) If F is the completion of E, then F is a Montel space.

Proof: Let u be the topology of E. By hypothesis the topology p on E' of uniform convergence on the precompact subsets of E is metrizable. Furthermore, p coincides with $\sigma(E',E)$ on each equicontinuous subset of E', and p is finer than $\sigma(E',E)$. Now, let B be any $\sigma(E',E)$-closed, circled, convex and bounded set in E' and $\{x_n'; n \geq 1\}$ a sequence in B. Since $\{x_n'; n \geq 1\}$ is equicontinuous, it is p-relatively compact and therefore $(x_n'; n \geq 1)$ has a subsequence which converges under p to an element of B. This shows that B is compact and therefore p-separable. The same is true with respect to the coarser topology $\sigma(E',E)$. In particular, $\sigma(E',E)$-separability of B implies that B is equicontinuous so that $u = \beta(E,E') = \tau(E,E')$ which proves (i). Let G be the dual of E' under p and H the u-quasi-completion of E. Since the closed, circled and convex envelope in H of each precompact subset of E is compact, it follows that $G \subset H$. It follows from Remark 3, that (E',p) is a Fréchet space so that $(G,\tau(G,E'))$ is complete ([43], Chapter 6, Corollary 2 of Proposition 1). Now $\sigma(E',E) \subset \sigma(E',G) \subset p$. Thus $\sigma(E'G)$ and $\sigma(E',E)$ coincide on each u-equicontinuous subset of E'. Since $u = \tau(E',E)$, it follows that $\tau(G,E')$ induces $\tau(E,E')$ on E and that G is the u-completion of E([43], Chapter 6, Theorem 3). Thus, $G = F = H$. Let now A be any $\sigma(F,E')$-bounded set. Then A is a pointwise bounded set of p-continuous linear functionals on the Fréchet space E'. By the Banach-Steinhaus theorem, A is equicontinuous and so there is a precompact subset C of E such that A is contained in the $\sigma(F,E')$-closed, circled and convex envelope of C. Thus A is a relatively compact subset of F. Since a $\sigma(E,E')$-bounded set is then relatively compact in F and hence precompact in E, (ii) now follows. It also follows that $\beta(E',E) = p$ so that F is the bidual of E which establishes (iii). (iv) is immediate, because the completion of a barrelled space is barrelled.

§3. Unconditional convergence and Bases.

As observed in Chapter 5, a topological vector space with a (countable) basis is separable and we have seen in Corollary 5 that a separable σ-barrelled space is barrelled. It is natural, therefore, to consider extended unconditional bases in σ-barrelled spaces. Most of the results of this section are due to Tweddle [51].

Theorem 5. Let E be a σ-barrelled space. Suppose that there is a family $\{x_\lambda\}_{\lambda\in\Lambda}$ of elements of E such that

(a) for each $x \in E$ there exist scalars α_λ, $\lambda \in \Lambda$ such that $\{\alpha_\lambda x_\lambda\}_{\lambda\in\Lambda}$ is unconditionally convergent to x and

(b) there is an element of E for which these scalars may all be chosen to be non-zero.

Then $(E,\tau(E,E'))$ is barrelled.

Proof: Let F be the completion of E and let F' be the $\tau(E',F)$-completion of E' and let v be the extension of u to F. Let B be a $\sigma(E',E)$-closed, circled, convex and bounded set. As in the proof of Proposition 4, it follows that B is $\sigma(E',F)$-countably compact so that, by Eberlein's theorem, B is $\sigma(F',F)$-relatively compact. Let y' be an element of the $\sigma(F',F)$-closure of B. Choose an element $z = \sum_{\lambda\varepsilon\Lambda} \beta_\lambda x_\lambda$ of E such that for all $\lambda \varepsilon \Lambda$, $\beta_\lambda \neq 0$ and let A be the v-closed, circled, convex envelope of $\{ \sum_{\lambda\varepsilon\phi} \beta_\lambda x_\lambda : \phi \varepsilon \Phi\}$, where Φ denotes the family of all finite subsets of Λ. By (Chapter I, §9, V), A is v-compact and therefore also $\sigma(F,E')$-compact. Now the same circled and convex subsets of F are compact under $\sigma(F,E')$ and $\sigma(F,F')$, so that A is $\sigma(F,F')$-compact ([43], Chapter 6, Corollary 4 of Theorem 2). Let M be the vector subspace of F generated by A and let $M' = F'/M^\circ$. Now $\langle M',M\rangle$ is a dual pair and $\sigma(F,F')$ induces $\sigma(M,M')$ on M. $(M',\tau(M',M))$ is, therefore, a normed space because A is a $\sigma(M,M')$-compact, circled, convex and absorbing subset of M. The quotient map $q: F' \to M'$ is clearly continuous under $\sigma(F',F)$ and $\sigma(M',M)$. Thus q(B) is $\sigma(M',M)$-countably compact and therefore $\sigma(M',M)$-compact, as in the proof of Proposition 4. In particular, q(B) is $\sigma(M',M)$-closed which implies that $q(y') \varepsilon q(B)$. Equivalently, there is an element $x' \varepsilon B$ such that $\langle x,x'\rangle = \langle x,y'\rangle$ for all $x \varepsilon M$. But $x_\lambda \varepsilon M$ for all $\lambda \varepsilon \Lambda$ because $\beta_\lambda x_\lambda \varepsilon A$ and $\beta_\lambda \neq 0$. Hence

$$\langle x_\lambda, x'\rangle = \langle x_\lambda, y'\rangle \tag{*}$$

for all $\lambda \varepsilon \Lambda$. Now let $x = \sum_{\lambda\varepsilon\Lambda} \alpha_\lambda x_\lambda \varepsilon E$. Repeating the argument applied above to z, it follows that y' is $\sigma(F,E')$-continuous on the u-closed, circled, convex envelope of $\{\sum_{\lambda\varepsilon\phi}\alpha_\lambda x_\lambda;\ \phi\ \varepsilon\ \Phi\}$. Thus by (*),

$$\langle x,y'\rangle = \sum_{\lambda\varepsilon\Lambda}\alpha_\lambda\langle x_\lambda y'\rangle = \sum_{\lambda\varepsilon\Lambda}\alpha_\lambda\langle x_\lambda x'\rangle = \langle x,x'\rangle$$

i.e. x' and y' coincide on E. Finally let $\bar{B}$ be the $\sigma(F',F)$-closure of B and let p be the quotient map of F' onto $F'/E°$, which contains E'. Then $p(\bar{B}) = B$ is $\sigma(F'/E°,E)$-compact and this completes the proof.

A particular case of Theorem 5 is the following:

<u>Corollary 8</u>. Let E be a countably barrelled space. Suppose that there is a family $\{x_\lambda\}_{\lambda\varepsilon\Lambda}$ of elements of E satisfying (a) and (b) of Theorem 5. Then $(E,\tau(E,E'))$ is barrelled.

The following example shows that condition (b) in Theorem 5 cannot be omitted.

<u>Example 3</u>. Let Λ be any uncountable set. Let E be the direct sum $R^{(\Lambda)}$ and E' the vector subspace of the product R^Λ consisting of all $x = \{x_\lambda\}_{\lambda\varepsilon\Lambda}$ for which at most countably many x_λ are non-zero; then $\langle E,E'\rangle$ is a dual pair. Given an atmost countable non-empty subset Γ of Λ and a family $\{\alpha_\lambda\}_{\lambda\varepsilon\Gamma}$ of real numbers, the set

$$\{\{x_\lambda\}_{\lambda\varepsilon\Lambda}:\ |x_\lambda| \le \alpha_\lambda,\ \lambda\ \varepsilon\ \Gamma\ ,\ x_\lambda = 0 \text{ otherwise}\}$$

is $\sigma(E',E)$-compact, circled and convex. The topology on E of uniform convergence on all such sets is a topology of the dual pair $\langle E,E'\rangle$ under which E is countably barrelled. $(E,\tau(E,E'))$ is not barrelled, because $E' \cap \prod_{\Lambda}[0,1]$ is $\sigma(E',E)$-closed and bounded but is not $\sigma(E',E)$-compact. If $x_\lambda = \{\delta_{\lambda\mu}\}_{\mu\varepsilon\Lambda}$, then clearly the family $\{x_\lambda\}_{\lambda\varepsilon\Lambda}$ satisfies (a) and not (b) in E.

It is interesting to note that E is semi-reflexive, so that a semi-reflexive countably barrelled space need not be barrelled.

Remark 4. If we assume in Theorem 5 that E is σ-quasibarrelled, then $(E,\tau(E,E'))$ is quasibarrelled.

We now show in the following theorem that the weak basis theorem generalizes, in the context of extended unconditional Schauder bases, to σ-barrelled spaces. For convenience an extended unconditional Schauder basis is referred to as e-basis.

Theorem 6. Let (E,u) be a σ-barrelled space and $\{x_\lambda\}_{\lambda\varepsilon\Lambda}$ an e-basis in E with respect to $\sigma(E,E')$. Then $\{x_\lambda\}_{\lambda\varepsilon\Lambda}$ is an e-basis for all topologies (up to $\beta(E,E')$ of the dual pair $\langle E,E'\rangle$. Also the family of coefficient functionals $\{x'_\lambda\}_{\lambda\varepsilon\Lambda}$ is an e-basis in E' for all topologies of the dual pair E',E .

Proof: Let Ω denote the set of finite subsets of $\mathbb{N}$. We show, first of all, that for each $x \varepsilon E$ and each countable subfamily $\{x_{\lambda(n)}\}_{n\varepsilon\mathbb{N}}$ of $\{x_\lambda\}_{\lambda\varepsilon\Lambda}$, the family $\{\langle x, x'_{\lambda(n)}\rangle x_{\lambda(n)}\}_{n\varepsilon\mathbb{N}}$ is unconditionally Cauchy with respect to $\tau(E,E')$. Suppose this is not the case. Then there exists an element $z \varepsilon E$, a

$\sigma(E,E')$-neighbourhood V of 0 and a countable subfamily $\{x_{\lambda(n)}\}_{n\in\mathbb{N}}$ such that for each $N \in \mathbb{N}$, there exists $\omega \in \Omega$ with $n > N$ for all $n \in \omega$ and $\sum_{n\in\omega} \langle z, x'_{\lambda(n)}\rangle x_{\lambda(n)} \notin V$. Proceeding inductively we can construct a sequence $\{\omega_m\}$ in Ω such that $\omega_r \cap \omega_s = \emptyset$ $(r \neq s)$ and for all m

$$(*) \qquad \sum_{n\in\omega_m} \langle z, x'_{\lambda(n)}\rangle x_{\lambda(n)} \notin V.$$

Consider the linear mapping T_m: $(E,u) \to (E,\tau(E,E'))$ defined, for each $m \in \mathbb{N}$, by

$$T_m(x) = \sum_{n\in\omega_m} \langle x, x'_{\lambda(n)}\rangle x_{\lambda(n)}.$$

If $x \in E$, $x' \in E'$ and T'_m is the transpose of T_m,

$$\langle x, T'_m(x')\rangle = \langle T_m(x), x'\rangle = \sum_{n\in\omega_m} \langle x, x'_{\lambda(n)}\rangle\langle x_{\lambda(n)}, x'\rangle$$

i.e. $\quad T'_m(x') = \sum_{n\in\omega_m} \langle x_{\lambda(n)}, x'\rangle x'_{\lambda(n)} \in E'.$

Thus, T'_m: $E' \to E'$ is $\sigma(E',E)$-continuous.

Let W be any closed, circled and convex $\tau(E,E')$-neighbourhood of 0. $T'_m(W^\circ)$ is $\sigma(E',E)$-compact and is contained in a finite dimensional vector subspace of E'. It is, therefore, $\sigma(E',E)$-separable which implies that $\bigcup_{m=1}^{\infty} T'_m(W^\circ)$ is $\sigma(E',E)$-separable.

By (Chapter 1, §9, V), for each $x \in E$, the set $\{T_m(x)\colon m \in \mathbb{N}\}$ is bounded, and so there exists $\gamma_{(x)} > 0$ such that $\{T_m(x)\colon m \in \mathbb{N}\} \subset \gamma_{(x)}W$. Then if $x' \in W^\circ$, $|\langle x, T'_m(x')\rangle| \leq \gamma_{(x)}$ for each m, which shows that $\bigcup_{m=1}^{\infty} T'_m(W^\circ)$ is $\sigma(E',E)$-bounded. Since it contains a $\sigma(E',E)$-dense bounded sequence, it must be u-equicontinuous. But

$$\left(\bigcup_{m=1}^{\infty} T'_m(W^\circ)\right)^\circ = \bigcap_{m=1}^{\infty}(T'_m(W^\circ))^\circ = \bigcap_{m=1}^{\infty} T_m^{-1}(W^{\circ\circ}) = \bigcap_{m=1}^{\infty} T_m^{-1}(W),$$

which shows that $\{T_m : m \in \mathbb{N}\}$ is equicontinuous for the stated topologies. There is, therefore, a u-neighbourhood U of 0 such that $T_m(U) \subseteq V$ for all m. Now the vector subspace of E generated by $\{x_\lambda\}_{\lambda\in\Lambda}$ is $\sigma(E,E')$-dense and therefore u-dense in E. Thus, there exists a finite subset ϕ of Λ and scalars α_λ, $(\lambda \in \phi)$ such that $z - \sum_{\lambda\in\phi} \alpha_\lambda x_\lambda \in U$. Also there is a positive integer N such that

$$\phi \cap \{\lambda(n) : n \in \bigcup_{m=N}^{\infty} \omega_m\} = \phi.$$

Then if $m \geq N$, $T_m(\sum_{\lambda\in\phi} \alpha_\lambda x_\lambda) = 0$ so that $T_m(z) \in T_m(U) \subseteq V$ which contradicts (*). It now follows that if F is the $\tau(E,E')$-completion of E and $x \in E$, $\Sigma\langle x, x'_{\lambda(n)}\rangle x_{\lambda(n)}$ is convergent in F for each countable subset $\{\lambda(n) : n \in \mathbb{N}\}$ of Λ. By (Chapter 1, §9, VI), $\{\langle x, x'_\lambda\rangle x_\lambda\}_{\lambda\in\Lambda}$ is, therefore, unconditionally Cauchy under $\tau(F,E')$ which induces $\tau(E,E')$ on E. Since $\{\langle x, x'_\lambda\rangle x_\lambda\}_{\lambda\in\Lambda}$ converges unconditionally to x under $\sigma(E,E')$, it must then converge likewise under $\tau(E,E')$, and hence under all topologies of the dual pair $\langle E,E'\rangle$. This proves the first assertion. To establish the other assertion, we proceed as follows: If $x \in E$, $x' \in E'$,

$$\langle x,x'\rangle = \left\langle \sum_{\lambda\in\Lambda} \langle x,x'_\lambda\rangle x_\lambda, x'\right\rangle = \sum_{\lambda\in\Lambda} \langle x,x'_\lambda\rangle\langle x_\lambda, x'\rangle$$

from which it follows that $\{\langle x_\lambda, x'\rangle x'_\lambda\}_{\lambda\in\Lambda}$ is unconditionally convergent to x' under $\sigma(E',E)$. Since E' is clearly $\sigma(E',E)$-sequentially complete, it follows from (Chapter I, §9, I, VI) that $\sigma(E',E)$ may be replaced by any topology of the dual pair $\langle E',E\rangle$. Finally, if $x' = \sum_{\lambda\epsilon\Lambda} \beta_\lambda x'_\lambda$ under $\sigma(E',E)$, for each $\mu\ \epsilon\Lambda$,

$$\langle x_\mu, x'\rangle = \sum_{\lambda\epsilon\Lambda} \beta_\lambda \langle x_\mu, x'_\lambda\rangle = \beta_\mu,$$

which shows that the coefficients are uniquely determined. The co-efficient functionals are the x_λ $(\lambda\ \epsilon\ \Lambda)$, which are continuous under any topology of the dual pair $\langle E',E\rangle$. This completes the proof.

A particular case of Theorem 6 is the following:

<u>Corollary 9</u>. Let (E,u) be a countably barrelled space. If $\{x_\lambda\}_{\lambda\epsilon\Lambda}$ is an e-basis in E for $\sigma(E,E')$, then it is an e-basis for all topologies (up to $\beta(E,E')$) of the dual pair E,E'. Also the family of coefficient functionals $\{x'_\lambda\}_{\Lambda\epsilon\lambda}$ is an e-basis in E' for all topologies of the dual pair E',E.

The following example shows that, unlike the Fréchet space case, a countably barrelled and hence a σ-barrelled space can have extended unconditional bases which are not e-bases.

<u>Example 4</u>. Let E and E' be as in Example 3 and choose $\lambda_0'\ \epsilon\ \Lambda$. Define

$$y_\lambda = \{\xi_\mu\}_{\mu\epsilon\Lambda}\ \epsilon\ E$$

by

$$\xi_\mu = 1 \text{ if } \mu = \lambda \text{ or } \mu = \lambda_0$$
$$= 0, \text{ otherwise.}$$

Then $\{y_\lambda\}_{\lambda\epsilon\Lambda}$ is an extended unconditional basis in E. But if

$x = \{x_\lambda\}_{\lambda \varepsilon \Lambda} \varepsilon E$,

$$x = \sum_{\lambda \neq \lambda_0} x_\lambda y_\lambda + (x_{\lambda_0} - \sum_{\lambda \neq \lambda_0} x_\lambda) y_{\lambda_0},$$

and the coefficient functional of y_{λ_0} clearly does not belong to E'.

§4. Sequentially barrelled spaces.

Husain [13], De Wilde and Houet [7] and Levin and Saxon [32] were all interested in the generalization of barrelled spaces and their interest led to the concepts and results which we discussed in Chapter 5 and in the preceeding sections of this present chapter. Likewise, while studying sequential convergence in locally convex spaces, Webb [56] was also interested in a generalization of barrelled spaces and he introduced the concept of sequentially barrelled space [56]. But it turns out, as we will show below, that sequentially barrelled spaces, in fact, form a proper generalization of σ-barrelled spaces. Most of the results of this section are due to Webb [56], [57] and [58].

Definition 4. An l.c. space (E,u) is said to be sequentially barrelled if each $\sigma(E',E)$-null sequence in E' is equicontinuous. Equivalently (E,u) is sequentially barrelled if each $\sigma(E',E)$-convergent sequence in E' is equicontinuous.

Proposition 6. Every σ-barrelled space is sequentially barrelled, but not conversely.

Proof: Let E be a σ-barrelled space and $\{f_n : n \in \mathbb{N}\}$ be a $\sigma(E',E)$-convergent sequence in E'. Then clearly $\{f_n : n \in \mathbb{N}\}$ is $\sigma(E',E)$-bounded and hence equicontinuous by definition.

For the converse we shall give counterexamples later.

We now prove a result which we have promised in Chapter 5 before Lemma 1.

Theorem 7. Let E be a sequentially barrelled space. Then

(a) each $\sigma(E',E)$-bounded subset of E' is $\beta(E',E)$-bounded.

(b) E is σ-barrelled iff E is σ-quasibarrelled.

(c) E is countably barrelled iff E is countably quasibarrelled.

(d) E is barrelled iff E is quasibarrelled.

Proof: (a) Let A and B be weakly bounded subsets of E and E' respectively. It is enough to show that $\sup\{|\langle x,y\rangle| ; x \in A, y \in B\} < \infty$. Suppose that $\sup\{|\langle x,y\rangle| : x \in A, y \in B\} = \infty$. Then there exists a sequence of points $\{x_n\}$ in B such that $\{\sup|\langle x,y_n\rangle| : x \in A\} > n^2$ for each n. Now $\{(1/n\, y_n)\}$ is $\sigma(E',E)$-null sequence and hence equicontinuous. This implies that $\{(1/n\, y_n)\}$ is strongly bounded. But $\sup\{|\langle x, 1/n\, y_n\rangle| : x \in A, n \in \mathbb{N}\} = \infty$, which is a contradiction. This completes the proof of (a).
(b), (c) and (d) follow easily from (a).

A particular case of Theorem 7 is the following:

Corollary 10. Let E be a σ-barrelled space. Then

(a) each $\sigma(E',E)$-bounded subset of E' is $\beta(E',E)$-bounded.

(b) E is countably barrelled iff E is countably quasibarrelled.

(c) E is barrelled iff E is quasibarrelled.

Corollary 11. Let E be a sequentially barrelled (DF)-space and M any countable codimensional subspace of E. Then M is a (DF)-space in the relative topology.

Proof. In view of Theorem 7 (c), and (Chapter 5, Remark 1), E is countably barrelled. But then, M is countably barrelled in the relative topology, by Proposition 11 of Chapter 5. Furthermore, since E has a fundamental sequence of bounded sets and since M is a subspace of E, it follows that M contains a fundamental sequence of bounded subsets of M. Hence M is a (DF)-space in the relative topology.

Notations. (i) If $\mathcal{F}_1$ and $\mathcal{F}_2$ are two families of subsets of E, we write $\mathcal{F}_1 \leq \mathcal{F}_2$ if for each $F_1 \in \mathcal{F}_1$ there is an $F_2 \in \mathcal{F}_2$ such that $F_1 \subseteq F_2$.

(ii) We write $\mathcal{N}$ to denote the class of u-null sequences in (E,u).

(iii) We write $\mathcal{K}_\sigma$ to denote the class of all circled, convex and $\sigma(E,E')$-compact subsets of (E,u).

(iv) Let (E,u) be an l.c. space. We write u^n to denote the topology on $E^{(b)}$ of uniform convergence on $\mathcal{N}$, where $E^{(b)}$ is the space of bounded linear functionals on E.

We need the following result, due to Grothendieck [10] :

Lemma 5. Let (E,u) be an l.c. space and K a subset of E^+. Then K is u-limited iff it is u^n-compact.

Theorem 8. An l.c. space (E,u) is sequentially barrelled iff $[\sigma(E',E)]^n \subseteq u$.

Proof: If (E,u) is sequentially barrelled, then clearly $[\sigma(E',E)]^n \subseteq u$. Conversely, assume the condition. Let $H = \{f_n;\ n = 1,2,\ldots\}$ be a $\sigma(E',E)$-null sequence in E'. Then H° is a $[\sigma(E',E)]^n$-neighbourhood of 0 and hence u-neighbourhood of 0 in E. Then $H^{\circ\circ}$ and hence H is equicontinuous. This completes the proof.

Corollary 12. Every precompact subset of a sequentially barrelled space (E,u) is $\sigma(E',E)$-limited.

Proof: If B is an precompact subset of (E,u) then clearly it is $[\sigma(E',E)]^n$-precompact. But then, B is $\sigma(E',E)$-limited in view of Lemma 5.

Proposition 7. (a) If (E,u) is a sequentially complete l.c. space, then $(E',\tau(E',E))$ is sequentially barrelled. (b) If (E,u) is a metrizable l.c. space such that $(E',\tau(E',E))$ is sequentially barrelled, then (E,u) is complete.

Proof: (a) Since (E,u) is sequentially complete, it follows from ([31], §20, 9(6)) that $\mathcal{N} \subset \mathcal{K}_\sigma$ and hence we conclude that $(E',\tau(E',E))$ is sequentially barrelled.

(b) Let $\{x_n\}$ be a Cauchy sequence in (E,u). Then $\{x_n\}$ is a u-precompact set so that there exists an u-null sequence $\{z_n\}$ in E such that $\{x_n\} \subseteq \overline{\Gamma}\{z_n\}$ ([31], §21, 10(3)). Since $(E',\tau(E',E))$ is sequentially barrelled, $\overline{\Gamma}\{z_n\}$ is $\sigma(E,E')$-compact, which implies that $\{x_n\}$ has a $\sigma(E,E')$-cluster point. This implies that (E,u) is complete.

Example 5. A Köthe function space Λ is sequentially barrelled in its Mackey topology $\tau(\Lambda,\Lambda^{\times})$: hence the same is true of a perfect sequence space ([56], page 354).

Example 6. $(\ell^{\infty},\tau(\ell^{\infty},\ell^{1}))$ is a separable sequentially barrelled space, with property (S), which is not σ-barrelled. We have seen in Example 1 that $(\ell^{\infty},\tau(\ell^{\infty},\ell^{1}))$ does not have property (C).

Example 7. Let c_0 be the space of all null sequences. Then $(\ell^{1},\tau(\ell^{1},c_0))$ is a sequentially barrelled space, without property (S), which is not σ-barrelled.

Example 8. $(c_0,\sigma(c_0,\ell^{1}))$ is a non-metrizable l.c. space which is not complete, although $(\ell^{1},\tau(\ell^{1},c_0))$ is sequentially barrelled. Thus part (b) of Proposition 7 may be false for non-metrizable l.c. spaces.

Remark 5. It is now clear from Examples 6 and 7 that a separable sequentially barrelled space need not be barrelled and that a sequentially barrelled space need not have properties (S) and (C), unlike σ-barrelled case.

Theorem 9. Let E be an l.c. space such that $(E,\gamma(E,E'))$ is complete, where $\gamma(E,E')$ is the topology on E of uniform convergence on the $\beta(E',E)$-bounded subsets of E'. Then $(E',\tau(E',E))$ is sequentially barrelled.

Proof: Let $\{x_n\}$ be a $\sigma(E,E')$-null sequence in E. We show that $\bar{\Gamma}\{x_n\}$ is $\sigma(E,E')$-compact and this will complete the proof. Let M be the $\gamma(E,E')$-closed linear span of $\{x_n\}$, equipped with the topology induced by $\gamma(E,E')$. Then M is a complete separable space and $\{x_n\}$ is bounded in M. We now construct a map T: $\ell^1 \to M$. Let $\eta = \{\eta_i\} \in \ell^1$. For $f' \in M'$, the sum $\langle f_\eta, f' \rangle = \sum_{i=1}^{\infty} \eta_i \langle x_i, f' \rangle$ is defined and so we see that $f_\eta \in (M')^*$. In order to show that $f_\eta \in M$, it is sufficient, in view of ([31], §21, 9(5)) to show that f_η is $\sigma(M',M)$-sequentially continuous on M'. So, let $\{f'_n\}$ be a $\sigma(M',M)$-null sequence in M' and $\varepsilon > 0$. Since M is complete, every $\sigma(M,M')$-bounded subset of M is $\beta(M,M')$-bounded. Hence

$$K = \sup\{|\langle x_i, f'_n \rangle|,\ 1 \leq i,\ n < \infty\} < \infty.$$

Choose j such that

$$\sum_{i=j+1}^{\infty} |\eta_i| < \frac{\varepsilon}{2(k+1)}.$$

Then choose N such that

$$|\eta_i \langle x_i, f'_n \rangle| < \frac{\varepsilon}{2j} \text{ for all } n > N \text{ and each } i \in [1,j].$$

Then

$|\langle f_\eta, f'_n \rangle| < \varepsilon$ whenever $n > N$. So, we have constructed a map of ℓ^1 into M which we denote by T, where

$$T\eta = f_\eta.$$

Clearly T is linear. Furthermore, it is continuous with respect to the topologies $\sigma(\ell^1, c_o)$ and $\sigma(M,M')$. Let B be the unit ball of ℓ^1. Then B is $\sigma(\ell^1, c_0)$-compact so that T(B) is $\sigma(M,M')$-compact in M. Let e_n denote the nth unit vector in ℓ^1. Then $Te_n = x_n$, and hence $\overline{\Gamma}\{x_n\} \subset T(B)$ which implies that $\overline{\Gamma}\{x_n\}$ is $\sigma(E,E')$-compact.

An immediate consequence of Theorem 9 is the following:

Corollary 13. If $(E,\gamma(E,E'))$ is complete, then every $\sigma(E,E')$-bounded subset of E is $\beta(E,E')$-bounded.

Proposition 8. Let E be an l.c. space such that $(E,\gamma(E,E'))$ is separable and $(E,\tau(E,E'))$ is sequentially barrelled. Then $(E,\tau(E,E'))$ and $(E,\gamma(E,E'))$ have the same completions.

Proof: Let f be in the completion of $(E,\tau(E,E'))$. Then f is $\sigma(E',E)$-continuous on every circled, convex, $\sigma(E',E)$-compact subset of E'. Since the $\sigma(E',E)$-convergent sequences in E' are $\tau(E,E')$-equicontinuous, f is $\sigma(E',E)$-sequentially continuous on E'. Now the separability of $(E,\gamma(E,E'))$ implies that the $\beta(E',E)$-bounded subsets of E' are $\sigma(E',E)$-metrizable. It follows that f is $\sigma(E',E)$-continuous on every $\beta(E',E)$-bounded subset of E', hence f is in the completion of $(E,\gamma(E,E'))$. The opposite inclusion follows easily, since $\tau(E,E') \leq \gamma(E,E')$.

Corollary 14. If $(E,\gamma(E,E'))$ is complete and separable and $(E',\beta(E',E))$ is complete, then $(E,\tau(E,E'))$ is complete.

Proof: Since $(E,\gamma(E,E'))$ is complete, $\gamma(E,E') = \beta(E',E)$ by Corollary 13. Applying Theorem 9 to $(E',\gamma(E',E))$ we deduce that $(E,\tau(E,E'))$ is sequentially barrelled. Hence Proposition 8 applies. This completes the proof.

Example. We give an example of a σ-barrelled space which is not countably barrelled. This example is due to I. Tweddle (unpublished) and is a modification of an example due to J. Schmets (Espaces associés à un espace linéaire à semi-normes- applications aux espaces de fonctions continues, Université de Liège, 1972-73), who in turn uses a modification of an example due to Levin and Saxon [32].

Let $F = \prod_{\mathbb{R}} \mathbb{R} = \mathbb{R}^{\mathbb{R}}$ ($\mathbb{R}$, real line), be the product space and $E = \Sigma_{\mathbb{R}} \mathbb{R}$, the direct sum. Then E and $E' = \{(x_\alpha)_{x \in \mathbb{R}} \in F: |\{\alpha; x_\alpha \neq 0\}| \leq \aleph_0\}$ are dual pairs. If B is a nonempty bounded subset of $\mathbb{R}$, then

$$S(B) = \{(x_\alpha)_{\alpha \in \mathbb{R}} \in E': \sum_{\alpha \in B} |x_\alpha| \leq 1 \text{ and } x_\alpha = 0 \text{ if } \alpha \notin B\}$$

is easily seen to be a $\sigma(E',E)$-compact absolutely convex subset.

Let ϕ be a countable set of $\mathbb{R}$ and for each $\alpha \in \phi$, let β_α be a nonnegative real number. Then $C = \prod_{\alpha \in \phi} [-\beta_\alpha, \beta_\alpha] \times \prod_{\alpha \notin \phi} \{0\}$ is a $\sigma(E',E)$-compact absolutely convex set. Let $\mathcal{A}$ be the collection of all such C and $S(B)$ where B is compact. By the Mackey-Arens theorem, the topology τ_0 on E defined by polars A°, where $A \in \mathcal{A}$, is compatible with duality between E and E'. If $\{x_\alpha^{(n)}\}_{\alpha \in \mathbb{R}}$ is a $\sigma(E',E)$-bounded sequence, for each integer n there is a countable subset ϕ_n of $\mathbb{R}$ such that $\beta_\alpha^{(n)} = 0$ for $\alpha \notin \phi_n$ and $\phi = \bigcup_{n=1}^{\infty} \phi_n$ is countable. By the boundedness, for each $\alpha \in \phi$,

$$\beta_\alpha = \sup\{|x_\alpha^{(n)}|: \; n \geq 1\} < \infty,$$

and so

$$\{\beta_\alpha^{(n)}\}_{\alpha\in\mathbb{R}} \in \{(x_\alpha)_{\alpha\in\mathbb{R}} \in E': |x_\alpha| \leq \beta_\alpha \text{ for } \alpha \in \phi \text{ and}$$

$$x_\alpha = 0 \text{ otherwise}\}.$$ This shows that (E,τ) is σ-barrelled. But we show that it is not countably barrelled. Clearly for any positive integer n, $[-n,n]$ is a compact subset of $\mathbb{R}$ and

$$S(\mathbb{R}) = \mathrm{Cl}\bigcup_{n=1}^{\infty} S[-n,n].$$

By the definition of τ_0, $S[-n,n]$ is τ_0-equicontinuous. We show that $S(\mathbb{R})$ is not τ_0-equicontinuous. Clearly a finite union of sets $C \in \mathcal{A}$ is also in $\mathcal{A}$ and

$$\bigcup_{n=1}^{k} S[B_n] \subset S[\bigcup_{n=1}^{k} B_n].$$

Hence every τ_0-neighbourhood of the origin contains a τ_0-neighbourhood of the form:

$$V = \varepsilon\{\{(x_\alpha)_{\alpha\in\mathbb{R}} \in F: |x_\alpha| \leq \beta_\alpha \; (\alpha \in \phi), \; x_\alpha = 0, \; \alpha \notin \phi\}^\circ \cap [S(B)]^\circ\}$$

where $\varepsilon > 0$, ϕ is countable, $\beta_\alpha > 0$, $\alpha \in \phi$ and B is a compact subset of $\mathbb{R}$. Choose $\alpha_0 \in \mathbb{R}\setminus(\phi \cup B)(\neq \emptyset)$, then if $x_\alpha = 1$ or 0 according as $\alpha = \alpha_0$ or otherwise, and $\eta_\alpha = 2$ or 0 according as $\alpha = \alpha_0$ or otherwise, we see that $(x_\alpha) \in S(\mathbb{R})$ but $(x_\alpha) \notin V^\circ$ because $(\eta_\alpha) \in V$ i.e. $S(\mathbb{R})$ is not contained in the polar of any τ_0-neighbourhood of 0.

§5. Open Problems.

Problem 1. (Levin and Saxon [32]). Is a Mackey space with property (C) always a σ-barrelled space?

Problem 2. Is $(E,\tau(E,E'))$ countably barrelled where E is the space of the above example and $\tau(E,E')$ is the Mackey topology?

Problem 3. If F is a σ-barrelled space, is it necessarily countably barrelled under the topology of uniform convergence on the $\sigma(F',F)$-bounded separable sets?

CHAPTER VII

COUNTABLY ULTRABARRELLED AND COUNTABLY QUASIULTRABARRELLED SPACES

The concepts of barrelled and quasibarrelled spaces led W. Robertson [44] and S.O. Iyahen [19] to introduce, respectively, the concepts of ultrabarrelled and quasiultrabarrelled spaces in situations where local convexity is not assumed. Similarly the concepts of countably barrelled and countably quasibarrelled spaces led Iyahen [20] to introduce the concepts of countably ultrabarrelled and countably quasiultrabarrelled spaces which we discuss in this chapter. Most of the results of this chapter are based on the work of Iyahen [20].

§1. Definitions and some properties.

Definition 1. Let E be a t.v. space. For each integer $k > 0$, let $\{V_k^{(n)}: n = 0,1,2,\ldots\}$ be a sequence of closed circled **0**-neighbourhoods in E such that $V_k^{(n+1)} + V_k^{(n+1)} \subset V_k^{(n)}$ for all n. If for each n, $V^{(n)} = \bigcap_{k=1}^{\infty} V_k^{(n)}$ is absorbing (bornivorous) then $V^{(0)}$ is clearly an ultrabarrel (a bornivorous ultrabarrel) in E with $\{V^{(n)}: n = 1,2,\ldots\}$ as a defining sequence. $V^{(0)}$ is called an ultrabarrel (a bornivorous ultrabarrel) of type (α).

Example 1. Any closed circled neighbourhood $V^{(0)}$ in a t.v.

space is a bornivorous ultrabarrel of type (α). For, there is a sequence $\{V^{(n)}: n = 1,2,\ldots\}$ of closed circled 0-neighbourhoods such that $V^{(n+1)} + V^{(n+1)} \subset V^{(n)}$ for all n. If for all $k \geq 1$ and each fixed n, $V_k^{(n)} = V^{(n)}$, then $\{V^{(n)}: n = 1,2,\ldots\}$ is a defining sequence for $V^{(0)}$.

Example 2. If $\{V_k^{(0)}: k \geq 1\}$ is a sequence of closed circled convex 0-neighbourhoods in a t.v. space, then $V^{(0)} = \bigcap_{k=1}^{\infty} V_k^{(0)}$ is an ultrabarrel (a bornivorous ultrabarrel) of type (α) whenever $V^{(0)}$ is absorbing (bornivorous) with $\{V^{(n)}: n \geq 1\}$, $V^{(n)} = \bigcap_{k=1}^{\infty} 2^{-n} V_k^{(0)}$, as a defining sequence.

Definition 2. A t.v. space E is called countably ultrabarrelled (countably quasiultrabarrelled) if each ultrabarrel (bornivorous ultrabarrel) of type (α) is a neighbourhood of 0 in E.

Proposition 1. (a) Every ultrabarrelled space is countably ultrabarrelled.

(b) Every quasiultrabarrelled space is countably quasiultrabarrelled.

(c) Every countably ultrabarrelled space is countably quasiultrabarrelled.

Proof: They are immediate in view of the definitions concerned.

Examples will be given later to show that the converse in each of (a), (b) and (c) in Proposition 1 is false.

Proposition 2. If (E,u) is a countably ultrabarrelled (countably quasiultrabarrelled) space, then $(E,u^{\circ\circ})$ is countably barrelled (countably quasibarrelled).

Proof: If $V^{(0)} = \bigcap_{k=1}^{\infty} V_k^{(0)}$ is an N-barrel in $(E,u^{\circ\circ})$, then $V^{(0)}$ is an ultrabarrel of type (α) in (E,u) and is therefore an u-neighbourhood of 0. The set $V^{(0)}$ must then be an $u^{\circ\circ}$-neighbourhood of 0 and thus $(E,u^{\circ\circ})$ is countably barrelled. Similarly, if (E,u) is countably quasibarrelled, then $(E,u^{\circ\circ})$ is countably quasibarrelled.

Corollary 1. Any l.c. countably ultrabarrelled (countably quasiultrabarrelled) space is countably barrelled (countably quasibarrelled), but not conversely.

Remark 1. For the converse in Corollary 1, a counterexample will be given later.

The following proposition is useful to demonstrate an example of a countably ultrabarrelled space which is not ultrabarrelled.

Proposition 3. The strong dual of a metrizable l.c. space is countably ultrabarrelled.

Proof: Let E be a metrizable l.c. space. Then $(E',\beta(E',E))$ is a complete l.c. space. It is enough to show that $(E',\beta(E',E))$ is countably quasiultrabarrelled, because by an argument similar to that in Proposition 4 (Chapter 3, §1) it will then be countably ultrabarrelled. Let $V^{(0)} = \bigcap_{k=1}^{\infty} V_k^{(0)}$ be a bornivorous ultrabarrel of type (α) with $\{V^{(n)} : n \geq 1\}$, $V^{(n)} = \bigcap_{k=1}^{\infty} V_k^{(n)}$, as a defining sequence. There exists, in $(E',\beta(E',E))$ a fundamental sequence say $\{B_n : n = 1,2,...\}$ of bounded sets which are circled, convex and weakly compact. Clearly for each n, there exists a positive real number λ_n such that $\lambda_n B_n \subseteq V^{(n+1)}$. Let A_n be the convex envelope of $\bigcup_{i=1}^{n} \lambda_i B_i$. Then A_n is a circled, convex and weakly compact subset of $\sum_{i=1}^{n} V^{(i+1)}$, and this is contained in $V^{(1)}$. For each i, there exists a weakly closed, circled, convex neighbourhood of 0 in $(E',\beta(E',E))$ such that $W_i \subset V_i^{(1)}$. The set $U_i = A_i + W_i$ is a weakly closed, circled, convex 0-neighbourhood in $(E',\beta(E',E))$ and $U_i \subseteq V_i^{(1)} + V^{(1)} \subseteq V_i^{(0)}$. The set $U = \bigcap_{i=1}^{\infty} U_i$ is weakly closed, circled and convex; it is also absorbing, since it absorbs each B_n. Moreover, $U \subseteq V^{(0)}$ which implies that $V^{(0)}$ is a neighbourhood of 0 since $(E',\beta(E',E))$ is countably barrelled. This completes the proof.

Example 3. We have shown in Proposition 3 that the strong dual of a metrizable l.c. space is countably ultrabarrelled. Köthe ([31],§31 (7)) has constructed a complete metrizable l.c. space whose strong dual is not quasibarrelled and hence, in view of Corollary 10, (Chapter 3, §1), it is not ultrabarrelled. Thus it is an example of a countably ultrabarrelled space which is not ultrabarrelled.

Example 4. We have seen in Example 1 (Chapter 2) that a normed space need not be countably barrelled, and hence, in view of Corollary 1, A normed space need not be countably ultrabarrelled. However, a normed space is quasiultrabarrelled and hence countably quasiultrabarrelled.

As we have seen in Chapter 3, the*-inductive limit topology does the job of the inductive limit topology in situations where local convexity is not assumed. Hence we have the following:

Theorem 1. The *-inductive limit of a family $\{[(E_\alpha, u_\alpha): f_\alpha],$ $\alpha \varepsilon A\}$ of countably ultrabarrelled (countably quasiultrabarrelled) spaces is again countably ultrabarrelled (countably quasiultrabarrelled).

Proof: Let (E,u) be the *-inductive limit of $\{[(E_\alpha, u_\alpha): f_\alpha]$ $\alpha \varepsilon A\}$. Let $V^{(0)}$ be an ultrabarrel of type (α) in (E,u) with $\{V^{(n)}: n = 1,2,\ldots\}$, $V^{(n)} = \bigcap_{i=1}^{\infty} V_i^{(n)}$, as a defining sequence. Clearly $f_\alpha^{-1}(V_i^{(n)})$ is a closed, circled neighbourhood of 0 in E_α for each α and i. Furthermore, it can easily be shown that $f_\alpha^{-1}(V^{(0)})$ is an ultrabarrel of type (α) in E_α with $\{f_\alpha^{-1}(V^{(n)}): n = 1,2,\ldots\}$ as a defining sequence. Hence $f_\alpha^{-1}(V^{(0)})$ and similarly each $f_\alpha^{-1}(V^{(n)})$ is a neighbourhood of 0 in E_α. This shows that each map $f_\alpha: E_\alpha \to (E,v)$ is continuous, where v is the linear topology on E with $\{V^{(n)}\}$ as a base of neighbourhoods of 0. Thus $v \subset u$ and hence $V^{(0)}$ is an u-neighbourhood of 0 in E. This shows that (E,u) is countably ultrabarrelled. Similarly, it can be shown that the *-inductive limit of a family of countably quasiultrabarrelled spaces is of the same sort.

An immediate consequence of Theorem 1 is the following:

Corollary 2. If E is a countably ultrabarrelled (countably quasiultrabarrelled) space and M a vector subspace of E, then E/M is countably ultrabarrelled (countably quasiultrabarrelled).

Proposition 4. The product of any family of countably ultrabarrelled (countably quasiultrabarrelled) spaces is again countably ultrabarrelled (countably quasiultrabarrelled).

Proof: Let $(E,u) = \prod_{\alpha\epsilon\phi} (E_\alpha,u_\alpha)$ be the product space of a family $\{(E_\alpha,u_\alpha)\}$ of countably ultrabarrelled spaces. If v_α is the finest linear topology on E_α, then (E_α,v_α) is ultrabarrelled and so $(E,v) = \prod_{\alpha\epsilon\phi}(E_\alpha,v_\alpha)$ is ultrabarrelled. If V is an ultrabarrel of type (α) in (E,u), then it is also an ultrabarrel of type (α) in (E,v) and is therefore a neighbourhood of 0 in (E,v). Thus for some finite subset ϕ_0 of ϕ, $\prod_{\alpha\epsilon\phi\setminus\phi_0} E_\alpha \subset V$.

Since a finite product of countably ultrabarrelled (countably quasiultrabarrelled) spaces has the same property, the result follows as in (Chapter 5, Proposition 10).

Proposition 5. Let (E,u) be a countably ultrabarrelled (countably quasiultrabarrelled) space and (F,v) any t.v. space. If f is a continuous, almost open and linear map of E into F, then (F,v) is countably ultrabarrelled (countably quasiultrabarrelled).

Proof: Let $V^{(0)}$ be an ultrabarrel (bornivorous ultrabarrel) of type (α) in (F,v) with $\{V^{(n)} : n = 1,2,\ldots\}$, $V^{(n)} = \bigcap_{i=1}^{\infty} V_i^{(n)}$, as a defining sequence. Then $f^{-1}(V^{(0)})$ is an ultrabarrel (bornivorous ultrabarrel) of type (α) in (E,u) with $\{f^{-1}(V^{(n)}):$ $n = 1,2,\ldots\}$, $f^{-1}(V^{(n)}) = \bigcap_{i=1}^{\infty} f^{-1}(V_i^{(n)})$, as a defining sequence. Hence $f^{-1}(V^{(0)})$ is a neighbourhood of 0 in (E,u). But f is almost open and $\overline{f(f^{-1}(V^{(0)}))} \subset V^{(0)}$, so that $V^{(0)}$ is a neighbourhood of 0 in (F,v). This completes the proof.

An example will be given in section 2 to show that the property of being countably ultrabarrelled (countably quasiultrabarrelled) is not inherited by closed vector subspaces.

§2. The Banach-Steinhaus and the isomorphism theorems.

Theorem 2. Let E be a countably ultrabarrelled (countably quasiultrabarrelled) space and F any topological vector space. If $\{H_n; n = 1,2,\ldots\}$, is a sequence of equicontinuous sets of linear mappings of E into F such that $\bigcup_{n\geq 1} H_n$ is pointwise bounded (respectively, uniformly bounded on bounded sets), then H is equicontinuous.

Proof: Let $V^{(0)}$ be a closed, circled neighbourhood of 0 in F.

Let $\{V^{(n)}: n = 1,2,\ldots\}$ be a sequence of closed circled neighbourhoods of 0 in F such that $V^{(n+1)} + V^{(n+1)} \subset V^{(n)}$ for all $n \geq 1$. Let $V_k^{(n)} = \bigcap_{f\varepsilon H_k} f^{-1}(V^{(n)})$, $k \geq 1$ and $W^{(n)} = \bigcap_{k=1}^{\infty} V_k^n = \bigcap_{f\varepsilon H} f^{-1}(V^{(n)})$. Then as in the proof of Proposition 7 of Chapter 5, it follows that $W^{(0)}$ is a neighbourhood of 0 in E and this is turn implies that H is equicontinuous.

Corollary 3. Let E and F be as in Theorem 2. If $\{f_n: n = 1,2,\ldots\}$ is a pointwise bounded (respectively, uniformly bounded on bounded sets) sequence of continuous linear maps of E into F, then it is equicontinuous.

Corollary 4. Let E be a countably ultrabarrelled space and F any topological vector space. Let $\{f_n; n = 1,2,\ldots\}$ be a sequence of continuous linear mappings of E into F, converging pointwise to a map f of E into F. Then f is continuous and linear and the convergence is uniform on precompact sets.

We have seen in Chapter 5 that the isomorphism theorem, which is valid for barrelled spaces, cannot be extended to countably barrelled spaces. However, the situation is different in non-locally convex case, and hence we have the following:

Theorem 3. Let E and F be countably ultrabarrelled spaces. Let $\{x_n,f_n\}$ and $\{y_n,g_n\}$ be Schauder bases in E and F respectively. Then, $\{x_n,f_n\}$ is similar to $\{y_n,g_n\}$ iff there exists an isomorphism

T of E onto F such that $Tx_n = y_n$ for all $n = 1,2,\ldots$.

Proof: The proof is practically same as that of Theorem 6 in Chapter 2 except that we have to use, now, Corollary 4 in place of Corollary 7 of Chapter 2.

As a particular case, we have the following:

Corollary 5. Let E and F be ultrabarrelled spaces and $\{x_n, f_n\}$ and $\{y_n, g_n\}$ Schauder bases in E and F respectively. Then $\{x_n, f_n\}$ is similar to $\{y_n, g_n\}$ iff there is an isomorphism T of E onto F such that $Tx_n = y_n$, $n = 1,2,\ldots$.

Proposition 5. Let E be a countably ultrabarrelled space and $\{x_i, f_i\}$ a Schauder basis in E. Then the sequence $\{T_n;\ n = 1,2,\ldots\}$ of maps $T_n\colon\ E \to E$, defined by $T_n(x) = \sum_{i=1}^{n} f_i(x)x_i$ is equicontinuous and converges uniformly on precompact sets to the identity mapping.

Proof: Clearly each T_n is continuous, linear and converges uniformly on compact sets to the identity mapping. Furthermore, $\{T_n\}$ is equicontinuous in view of Corollary 3.

Corollary 6. Let E be an ultrabarrelled space and $\{x_i, f_i\}$ a Schauder basis in E. Then the sequence $\{T_n\}$ of maps $T_n\colon\ E \to E$, defined by $T_n(x) = \sum_{i=1}^{n} f_i(x)x_i$ is equicontinuous and converges uniformly to the identity mapping on precompact sets.

Corollary 3 is now used to construct a counterexample promised in Remark 1.

Example 5. Consider $\ell^{\frac{1}{2}} = \{x = \{x_i\}: \sum_{i=1}^{\infty} |x_i|^{\frac{1}{2}} < \infty\}$ equipped with the topology u as defined in Chapter 3, §1, Example 2. As has been shown there $(\ell^{\frac{1}{2}},u^{\circ\circ})$ is barrelled and hence countably barrelled. However we show that $(\ell^{\frac{1}{2}},u^{\circ\circ})$ is not countably ultrabarrelled.

For $x = \{x_i\} \in \ell^{\frac{1}{2}}$, define

$$f_n(x) = (x_1,x_2,\ldots,x_n,0,0,\ldots).$$

Then $\{f_n;\ n = 1,2,\ldots\}$ is a sequence of continuous linear maps of $(\ell^{\frac{1}{2}},u^{\circ\circ})$ into $(\ell^{\frac{1}{2}},u)$, which is uniformly bounded on bounded sets. But since the identity mapping of $(\ell^{\frac{1}{2}},u^{\circ\circ})$ onto $(\ell^{\frac{1}{2}},u)$ is not continuous $\{f_n;\ n = 1,2,\ldots\}$ is not equicontinuous. Hence by Corollary 3 $(\ell^{\frac{1}{2}},u^{\circ\circ})$ is not countably quasiultrabarrelled. Observe that $(\ell^{\frac{1}{2}},u^{\circ\circ})$ is the inductive limit of Banach spaces. The space $(\ell^{\frac{1}{2}},u^{\circ\circ})$ being Hausdorff and complete, can be embedded as a closed vector subspace of a product $\wp$ of Banach spaces. Clearly $\wp$ is ultrabarrelled and hence countably ultrabarrelled. It follows that a closed vector subspace of a countably ultrabarrelled (countably quasiultrabarrelled) space need not be of the same sort.

§3. Tensor Products.

In this section we show that the (projective) tensor product of two metrizable countably ultrabarrelled spaces (in particular, metrizable ultrabarrelled spaces) is countably ultrabarrelled (respectively, ultrabarrelled). The result is well known for barrelled spaces, as we saw in Chapter 2. Its extension to metrizable countably barrelled spaces is vacuous, because each metrizable countably barrelled space is barrelled (Chapter 5, Corollary 5).

<u>Theorem 4</u>. Let G be a t.v. space and E, F any metrizable t.v. spaces such that E is countably ultrabarrelled. Let H be a family of mappings of $E \times F$ into G such that

(i) for each fixed $y \in F$, $\{f(.,y);\ f \in H\}$ is an equicontinuous subset of $\mathcal{L}(E,G)$;

(ii) for each fixed $x \in E$, $\{f(x,.);\ f \in H\}$ is an equicontinuous subset of $\mathcal{L}(F,G)$.

Then H is an equicontinuous family of mappings of $E \times F$ into G.

<u>Proof</u>: Clearly $E \times F$ is metrizable. Hence, in view of (Chapter 1, Lemma 1), it is sufficient to show that if $\{(x_n,y_n);\ n \geq 1\}$ is a sequence in $E \times F$ converging to $(x_0,y_0) \in E \times F$, then $\{f(x_n,y_n)\}$ converges to $f(x_0,y_0)$ uniformly in $f \in H$. Let $f_{y_n} = f(.,y_n)$. We show that $\{f_{y_n} - f_{y_0};\ n \geq 1,\ f \in H\}$ is pointwise bounded in $\mathcal{L}(E,G)$, i.e. for each $x \in E$, the set $N_x = \{f(x,y_n) - f(x,y_0),\ n \geq 1,\ f \in H\}$ is bounded in G. Now, if V is a circled neighbourhood of 0 in F, (ii) implies that there exists an integer n_0 such that for each $n \geq n_0$, $f(x,y_n) - f(x,y_0) \in V$ for each $f \in H$; on the other hand, for each $y \in F$, the set $\{f(x,y) = f_y(x);\ f \in H\}$ is bounded in G,

by (i) and (Chapter 1, Proposition 13) for each integer $n < n_0$, the set $\{f(x,y_n) - f(x,y_0)\}$ is bounded in F. There exists, therefore, a $\lambda \geq 1$ such that $f(x,y_n) - f(x,y_0) \in \lambda V$ for $n < n_0$ and for each $f \in H$. We now conclude that $N_x \subset \lambda V$ which proves that N_x is bounded. Since E is countably ultrabarrelled, $\{f_{y_n} - f_{y_0} ; n \geq 1, f \in H\} = \bigcup_{n=1}^{\infty} \{f_{y_n} - f_{y_0} ; f \in H\}$ is equicontinuous. Hence, in view of (i), $\{f_{y_n} ; n \geq 1, f \in H\}$ is equicontinuous. This implies that, for each neighbourhood V of 0 in G, there exists a neighbourhood U of x_0 in E such that $f(x,y_n) - f(x_0,y_n) \in V$ for each $x \in U$, each $n \geq 1$ and each $f \in H$. But if n is sufficiently large, $x_n \in U$ where $f(x_n,y_n) - f(x_0,y_n) \in V$. On the other hand it follows from (ii) that, if n is sufficiently large then $f(x_0,y_n) - f(x_0,y_0) \in V$ for each $f \in H$. There exists, then, an integer m such that, for each $n \geq m$ and each $f \in H$, $f(x_n,y_n) - f(x_0,y_0) \in V + V$ which implies that the sequence $\{f(x_n,y_n)\}$ converges to $f(x_0,y_0)$ uniformly in $f \in H$.

A particular case of Theorem 4 is the following:

Corollary 7. If E and F are metrizable t.v. spaces such that E is countably ultrabarrelled and if G is any t.v. space, then any separately continuous bilinear map of $E \times F$ into G is continuous.

Let E and F be any t.v. spaces. The canonical map $g\colon E \times F \to E \otimes F$ is continuous when $E \otimes F$ is equipped with the trivial topology. Therefore, if P is the set of all linear topologies on $E \otimes F$ for which g is continuous, then P is not empty. If u is the upper bound of P, then U is in P and is the finest linear topology on $E \otimes F$ for which g is continuous. The t.v. space $(E \otimes F, w)$ is called the topological tensor product of E and F.

Let $f: (E \otimes F, u) \to G$ be a linear map, where G is any t.v. space with η as a base of neighbourhoods of 0 for the topology of G. Then the family $\{f^{-1}(V); V \in \eta\}$ of sets is a base of neighbourhoods of 0 for a linear topology, say v, on $E \otimes F$. f is continuous iff $v \subset u$ and this is so iff $g: E \times F \to (E \otimes F, v)$ is continuous. Thus f is continuous iff $f \circ g: E \times F \to G$ is continuous and this implies that the algebraic isomorphism:

$$L((E \otimes F, u), G) \to L(E \times F, G),$$

is continuous. Let η and ζ be the bases of neighbourhoods of 0 for topological vector spaces E and F respectively. The family $\{\bigcup_{k=1}^{\infty} \sum_{k=1}^{n} g(U_k, V_k)\}$ of sets, as U_k and V_k vary over sequences from η and ζ respectively, is a base of neighbourhoods of 0 for a linear topology, say w, on $E \otimes F$.

<u>Lemma 1</u>. $w = u$.

<u>Proof</u>: Clearly $g: E \times F \to (E \otimes F, \omega)$ is continuous and so $w \subset u$. We now show that $u \subset w$. Let W be an u-neighbourhood of 0 and $\{W_n\}$ a sequence of u-neighbourhoods of 0 such that $W_1 + W_1 \subseteq W$ and $W_{n+1} + W_{n+1} \subseteq W_n$ for all $n \geq 1$. Since $g: E \times F \to (E \otimes F, u)$

is continuous, there are sequences $\{U_k\} \subset \eta$ and $\{V_k\} \subset \zeta$ such that

$$\bigcup_{n=1}^{\infty} \sum_{k=1}^{n} g(U_k, V_k) \subset \bigcup_{n=1}^{\infty} \sum_{k=1}^{n} W_k \subset W \text{ and hence}$$

$u \subset w$. This completes the proof.

Theorem 5. If E and F are any metrizable countably ultrabarrelled spaces, then $(E \otimes F, u)$ is countably ultrabarrelled.

Proof: Let $V^{(0)}$ be an ultrabarrel of type (α) in $(E \otimes F, u)$ with $\{V^{(n)};\ n = 1,2,\ldots\}$ as a defining sequence. Let λ be the linear topology on $E \otimes F$, with $\{V^{(n)}\}$ as a base of neighbourhoods of 0. Let $g\colon E \times F \to (E \otimes F, u)$ be the canonical map. Then g is continuous and so each partial map $g_{y_0}\colon E \to (E \otimes F, u)$, $y_0 \in F$ is continuous. Hence $g_{y_0}^{-1}(V^{(0)})$ is an ultrabarrel of type (α) in E and so $g_{y_0}^{-1}(V^{(0)})$ is a neighbourhood of 0. This implies that $g_{y_0}\colon E \times F \to (E \otimes F, v)$ is continuous. It follows easily from here that the map $g\colon E \times F \to (E \otimes F, \lambda)$ is separately continuous and so continuous in view of Corollary 7. Thus $\lambda \subset u$ and so $V^{(0)}$ is an u-neighbourhood of 0. This completes the proof.

Corollary 8. If E and F are metrizable ultrabarrelled spaces, then $(E \otimes F, u)$ is ultrabarrelled.

§4. Open Problem.

Problem 1. If the conclusion of Theorem 2 is true, is E a countably ultrabarrelled (countably quasiultrabarrelled) space?

CHAPTER VIII

COUNTABLY ORDER-QUASIULTRABARRELLED VECTOR LATTICES AND SPACES

In this chapter, we discuss a class of l.c. vector lattices (ordered l.c. spaces) which generalizes that of order-quasibarrelled vector lattices (respectively, order-quasibarrelled vector spaces). Also, analogous to the (DF)-spaces of Grothendieck, we introduce here, what we call order-(DF)-vector lattices and spaces. All the results of this chapter, except those on order-(DF)-vector lattices and spaces which are due to the second author, are based on the authors' paper [15].

§1. C.O.Q. Vector Lattices.

Definition 1. (a) Let (E,C,u) be an ordered l.c. space. An order-bornivorous barrel which is the countable intersection of closed, convex and circled neighbourhoods of 0 in E is called an order-bornivorous N-barrel.

(b) Let (E,C,u) be an l.c. vector lattice. A solid barrel which is the countable intersection of closed, convex and solid neighbourhoods of 0 in E is called a solid N-barrel.

Definition 2. Let (E,C,u) be an l.c. vector lattice. (E,C,u) is called a countably order-quasibarrelled vector lattice (abbreviated to C.O.Q. vector lattice) if each order-bornivorous

N-barrel is a neighbourhood of 0 in E.

Theorem 1. An l.c. vector lattice (E,C,u) is a C.O.Q. vector lattice iff each solid N-barrel is a neighbourhood of 0 in E.

Proof: The "only if" part follows easily, since a solid N-barrel is an order-bornivorous N-barrel. To establish the "if" part, let $B = \bigcap_{n\geq 1} V_n$ be an order-bornivorous N-barrel. Let $K(B)$ be the solid kernel of B. Then, it easily follows that

$$K(B) = \bigcap_{n\geq 1} K(V_n).$$

Since $K(V_n)$ is the largest solid set contained in V_n, and since (E,C,u) is an l.c. vector lattice, there exists a convex, solid neighbourhood U_n of 0 in (E,C,u) such that $U_n \subset K(V_n) \subset V_n$ and hence $K(V_n)$ is a neighbourhood of 0 in (E,C,u), for each $n \geq 1$. As shown in the proof of Theorem 1, Chapter 4, each $K(V_n)$ is closed and convex. Similarly $K(B)$ is closed and convex. Also $K(B)$ is order-bornivorous (Chapter 4, Lemma 1). Thus $K(B) = \bigcap_{n\geq 1} K(V_n)$ is a solid N-barrel in (E,C,u) and hence a neighbourhood of 0, by assumption. But then B is a neighbourhood of 0 in (E,C,u) because $K(B) \subset B$. This completes the proof.

Theorem 2. An l.c. vector lattice (E,C,u) is a C.O.Q. vector lattice iff each $\sigma_s(E',E)$-bounded subset of E' which is the countable union of equicontinuous subsets of E' is itself equicontinuous.

Proof: The proof can be accomplished as in (Chapter 5, Theorem 1), if we note that a subset H of E' is $\sigma_s(E',E)$-bounded iff H° is order-bornivorous.

Corollary 1. Every order-quasibarrelled vector lattice is a C.O.Q. vector lattice.

Examples will be given later to show that the converse in Corollary 1 is false.

Proposition 1. (a) A countably barrelled l.c. vector lattice is a C.O.Q. vector lattice;

(b) A C.O.Q. vector lattice is a countably quasibarrelled l.c. vector lattice.

Proof: (a) This is immediate in view of the definitions concerned.

(b) This, also, is immediate if we note that every bornivorous N-barrel in a countably quasibarrelled l.c. vector lattice is, a fortiori, an order-bornivorous N-barrel.

Let X be a completely regular Hausdorff space and C(X) the l.c. vector lattice of all continuous real-valued functions on X, ordered by the cone $K = \{f \in C(X);\ f(t) \geq 0 \text{ for all } t \in X\}$ and endowed with the compact-open topology u_c. Then we have the following result analogous to (Chapter 4, Proposition 8).

Proposition 2. The following statements are equivalent:

(a) $(C(X),K,u_c)$ is a countably barrelled l.c. vector lattice.

(b) $(C(X),K,u_c)$ is a C.O.Q. vector lattice.

(c) Every $C(X)$-pseudocompact subset of X which is the closure of a countable union of compact sets is actually compact.

<u>Proof</u>: That (a) implies (b) follows from Proposition 1, part (a). The equivalence of (a) and (c) has been established in (Chapter 5, Theorem 4). We now show that (b) implies (c) to complete the proof. As in (Chapter 4, Proposition 8), X is homeomorphic to $\hat{X} \subset (C(X)',\sigma(C(X)',C(X)))$ under the map $x \to \hat{x}$ defined by

$$\hat{x}(f) = f(x) \text{ for all } f \in C(X).$$

Let B be a $C(X)$-pseudocompact subset of X such that $B = cl(\bigcup_{n\geq 1} K_n)$, where each K_n is a compact subset of X. Define $\hat{A}_n = \{\hat{x} \in C(X)';\ \{x\} = supp(\hat{x}) \subset K_n$ and for every $f \in C(X)$ with $p_{K_n}(f) \leq 1\}$. Then by (Chapter 5, Lemma 2), $\hat{A}_n$ is equicontinuous; it now follows from $|\hat{x}|(|f|) = \sup\{|\hat{x}(g)|;\ |g| \leq |f|\}(f \in C(X))$ that $A = \bigcup_{n\geq 1} \hat{A}_n$ is a $\sigma_s(C(X)',C(X))$-bounded subset of $C(X)'$, because $B = cl(\bigcup_{n\geq 1} K_n)$ is $C(X)$-pseudocompact. Hence $\hat{A}$ is equicontinuous by assumption. But then by (Chapter 5, Lemma 2) $supp(\hat{A}) = B$ is a compact subset of X. This completes the proof.

The following examples show that the converse of Corollary 1 is false.

<u>Example 1</u>. $(C(W),K,u_c)$, where W is the space of ordinals less than the first uncountable ordinal is a C.O.Q. vector lattice by Proposition 2, but is not an order-quasibarrelled vector

lattice in view of (Chapter 4, Proposition 8).

Example 2. Let (E,C,u) be any metrizable l.c. vector lattice. Then (E',C',β(E',E)) is an l.c. vector lattice (Chapter 1, Proposition 41). But (E',β(E',E)) is countably barrelled (Chapter 5, Example 2) and so (E',C',β(E',E)) is a C.O.Q. vector lattice by Proposition 1, part (a).

Examples will be given later to show that the converse of each of the statements of Proposition 1 is false.

We now consider various conditions under which a C.O.Q. vector lattice is countably barrelled and a countably quasi-barrelled l.c. vector lattice is a C.O.Q. vector lattice.

Proposition 3. Let (E,C,u) be a C.O.Q. vector lattice such that C' is a strict $\mathcal{B}$-cone in (E',σ(E',E)). Then (E,C,u) is a countably barrelled l.c. vector lattice.

Proof: Let $B = \bigcap_{n \geq 1} V_n$ be an N-barrel in E. Then B° is a σ(E',E)-bounded subset of E'. But then, in view of (Chapter 4, Lemma 9) B° is a σ_s(E',E)-bounded subset of E'. This implies that B is an order-bornivorous N-barrel in E and hence a neighbourhood of 0.

Proposition 4. Let (E,C,u) be a C.O.Q. vector lattice satisfying condition Ⓐ: The topology β(E,E') on E is the relative topology induced by $\beta(E''_{|\sigma|}, E')$, and the σ(E',E)-closure of each σ_s(E',E)-bounded set in E' is σ_s(E',E)-bounded. Then (E,C,u) is a countably barrelled l.c. vector lattice.

Proof: The proof is immediate if we note that the condition Ⓐ implies that each $\sigma(E',E)$-bounded subset of E' is $\sigma_s(E',E)$-bounded (Chapter 4, Lemma 6).

In view of (Chapter 1, Proposition 38), (Chapter 5, Corollary 3) and Proposition 1, part (b), we have the following:

Proposition 5. If (E,C,u) is a C.O.Q. vector lattice, then the completion $(\tilde{E},\tilde{C})$ of (E,C) is a countably barrelled l.c. vector lattice for the order structure determined by the closure $\tilde{C}$ of C in $(\tilde{E},\tilde{C})$.

Also, in view of (Chapter 5, Proposition 2) and Proposition 1, part (b), the proof of the following proposition is immediate.

Proposition 6. A sequentially complete C.O.Q. vector lattice is a countably barrelled l.c. vector lattice.

Corollary 2. A quasi-complete (in particular, complete) C.O.Q. vector lattice is a countably barrelled l.c. vector lattice.

Now we consider conditions under which a countably quasibarrelled l.c. vector lattice is a C.O.Q. vector lattice.

Proposition 7. Let (E,C,u) be a countably quasibarrelled l.c. vector lattice such that E' is a normal subspace of E^b. Then (E,C,u) is a C.O.Q. vector lattice.

Proof: Let $\{H_n;\ n = 1,2,\ldots\}$ be a countable family of equicontinuous subsets of E' such that $H = \bigcup_{n\geq 1} H_n$ is $\sigma_s(E',E)$-bounded.

But then, as in the proof of Proposition 1, Chapter 4, H is $\beta(E',E)$-bounded and hence equicontinuous. This shows that (E,C,u) is a C.O.Q. vector lattice.

Corollary 3. Let (E,C,u) be a countably quasibarrelled l.c. vector lattice such that E' is $\sigma_s(E',E)$-complete. Then (E,C,u) is a C.O.Q. vector lattice.

Proof: We observe that E' is $\sigma_s(E',E)$-complete iff E' is a normal subspace of E^b, and hence Proposition 7 applies.

Proposition 8. Let (E,C,u) be a countably quasibarrelled l.c. vector lattice satisfying condition Ⓑ: The topology $\beta(E''_{|\sigma|},E')$ on $E''_{|\sigma|}$ is the relative topology induced by $\beta(E'',E')$ and the $\sigma_s(E',E)$-closure of each $\beta(E',E)$-bounded set in E' is $\beta(E',E)$-bounded. Then (E,C,u) is a C.O.Q. vector lattice.

Proof: This follows at once if we note that condition Ⓑ implies that each $\sigma_s(E',E)$-bounded set in E' is $\beta(E',E)$-bounded (Chapter 4, Lemma 5).

We have already seen (Examples 1 and 2) that C.O.Q. vector lattices form a proper generalization of order-quasibarrelled vector lattices. It is, therefore, interesting to find condition (or conditions) under which a C.O.Q. vector lattice is order-quasibarrelled.

The following proposition is useful in this direction; its proof is similar to that of Theorem 7, part (a) of Chapter 6 and hence omitted.

<u>Proposition 9</u>. Let (E,C,u) be a C.O.Q. vector lattice. Then each $\sigma_s(E',E)$-bounded subset of E' is $\beta(E',E)$-bounded.

<u>Corollary 4</u>. Let (E,C,u) be a C.O.Q. vector lattice. Then (E,C,u) is an order-quasibarrelled vector lattice iff it is a quasibarrelled l.c. vector lattice.

<u>Proof:</u> The "only if" part is obvious. The "if" part follows from Proposition 9.

<u>Corollary 5.</u> Every metrizable (or bornological) C.O.Q. vector lattice is an order-quasibarrelled vector lattice.

Now we give examples to show that the converse of each of the statements in Proposition 1 is false.

<u>Example 3</u>. Consider the Banach lattice m and the subspace E of m as in (Chapter 4, Example 1). As has been shown there, E is an order-quasibarrelled vector lattice and hence by Corollary 1 it is a C.O.Q. vector lattice. But E is not countably barrelled; for if it were, then by Corollary 5 of Chapter 5, it would be barrelled which is not true (Chapter 4, Example 1).

<u>Example 4</u>. Consider the normed vector lattice ϕ of all real sequences with only finitely many nonzero coordinates, equipped with the usual pointwise ordering and the supremum norm. Being normed, ϕ is clearly quasibarrelled and hence countably quasi-barrelled. But it is not a C.O.Q. vector lattice; for if it were, then by Corollary 5, it would be order-quasibarrelled which is not true (Chapter 4, Example 2).

§2. Order-(DF)-Vector lattices.

We now introduce a subclass of C.O.Q. vector lattices, which we call the class of order-(DF)-vector lattices. The notion of (DF)-space which is due to Grothendieck led us to introduce this special subclass of C.O.Q. vector lattices.

Definition 3. An ordered vector space (E,C) is said to possess a fundamental sequence of order-bounded sets, $\{B_n\}$ if every order-bounded set B in E is contained in some B_n.

Definition 4. A C.O.Q. vector lattice (E,C,u) with a fundamental sequence of order-bounded sets is called an order-(DF)-vector lattice.

Proposition 10. A C.O.Q. vector lattice with an order-unit is an order-(DF)-vector lattice.

Proof: Let e be an order-unit in a C.O.Q. vector lattice (E,C,u). Then the family $\{n[-e,e];\ n \geq 1\}$ forms a fundamental sequence of order-bounded sets in E. Hence (E,C,u) is an order-(DF)-vector lattice.

Corollary 6. Every order-unit-normed vector lattice is an order-(DF)-vector lattice.

Proof: We observe that an order-unit-normed-vector lattice is always an order-quasibarrelled vector lattice and hence a C.O.Q. vector lattice. Proposition 10 now applies.

Example 5. The Banach lattice m of all bounded real sequences with the usual pointwise ordering and the supremum norm is a C.O.Q. vector lattice. The element $e = \{e_n\}$ defined by $e_n = 1$ for all $n = 1,2,\ldots$ is an order-unit in m. Hence m is an order-(DF)-vector lattice.

Similarly the Banach lattice c of all convergent real sequences is an order-(DF)-vector lattice.

Example 6. The Banach lattice C(I) of all continuous real-valued functions on $I = [0,1]$ with the supremum norm and the order structure determined by the positive cone $K = \{f \in C(I);\ f(t) \geq 0 \text{ for all } t \in I\}$ is an order-(DF)-vector lattice, since the function identically equal to one on I is an order-unit in C(I).

The l.c. vector lattice $(C(W),K,u_c)$ of Example 1 is in fact an order-(DF)-vector lattice which is not order-quasibarrelled. The order-unit-normed vector lattice E of Example 3 is an order-(DF)-vector lattice which is not countably barrelled. The following example shows that an order-quasibarrelled vector lattice and hence a C.O.Q. vector lattice need not be an order-(DF)-vector lattice. It also shows that a countably barrelled l.c. vector lattice need not be an order-(DF)-vector lattice.

Example 7. Consider the Banach lattice $\ell^1 = \{x = \{x_i\};\ x_i \in \mathbb{R},\ \sum_{i=1}^{\infty} |x_i| < \infty\}$, equipped with the norm $||x|| = \sum_{i=1}^{\infty} |x_i|$ and ordered by the positive cone $K = \{x = \{x_i\} \in \ell^1; x_i \geq 0$ for all $i \geq 1\}$: Clearly $(\ell^1, K, \|\cdot\|)$ is an order-quasibarrelled vector lattice as well as a countably barrelled l.c. vector lattice. Hence it is a C.O.Q. vector lattice. We show, however, that it is not an order-(DF)-vector lattice. To show this, we have to establish that ℓ^1 does not have a fundamental sequence of order-bounded sets. Suppose it has. Then $\sigma_s(\ell^\infty, \ell^1)$ is metrizable. Since $(\ell^\infty, \sigma_s(\ell^\infty, \ell^1))$ is complete, it follows that $(\ell^\infty, \sigma_s(\ell^\infty, \ell^1))$ is complete metrizable and hence $\sigma_s(\ell^\infty, \ell^1)$ coincides with the supremum norm topology of ℓ^∞, because the two topologies are comparable. Thus, $\sigma_s(\ell^\infty, \ell^1)$ is normable. Clearly the positive cone K in ℓ^1 is generating and $\sigma(\ell^1, \ell^\infty)$-closed. This implies that ℓ^1 has an order-unit. But we know that ℓ^1 does not have an order-unit and thus a contradiction.

<u>Proposition 12</u>. Let (E,C,u) be an order-(DF)-vector lattice. Then

(a) $E'_{|\sigma|} = (E', \sigma_s(E',E))$ is a metrizable l.c. vector lattice.

(b) The strong dual of $E'_{|\sigma|}$ is an l.c. vector lattice which is a complete (DF)-space.

<u>Proof</u>: (a) If $\{B_n; n \geq 1\}$ is a fundamental sequence of order-bounded sets in E, then the sequence $\{B_n^\circ: n \geq 1\}$ forms a neighbourhood basis of 0 for $\sigma_s(E',E)$, which shows that $E'_{|\sigma|}$ is metrizable.

(b) This follows easily in view of (Chapter 1, Proposition 41) and (Chapter 2, Remark before Theorem 16).

Corollary 7. The strong dual of $E'_{|\sigma|}$ is a countably barrelled l.c. vector lattice.

§3. Permanence properties.

Theorem 3. The locally convex direct sum of a family of C.O.Q. vector lattices is a C.O.Q. vector lattice.

Proof: Let $\{(E_\alpha, C_\alpha, u_\alpha);\ \alpha \in I\}$ be a family of C.O.Q. vector lattices. Let $E = \oplus_\alpha E_\alpha$, $C = \oplus_\alpha C_\alpha$, $u = \oplus_\alpha u_\alpha$ and $i_\alpha : E_\alpha \to E$ the injection map for each $\alpha \in I$. Then (E,C,u) is an l.c. vector lattice (Chapter 1, Proposition 36(b)). We show that (E,C,u) is a C.O.Q. vector lattice. Let $B = \bigcap_{n \geq 1} V_n$ be a solid N-barrel in E. Clearly $i_\alpha^{-1}(B) = \bigcap_{n \geq 1} i^{-1}(V_n)$. Since each i_α is a continuous lattice homomorphism of E_α into E, it follows that $i_\alpha^{-1}(V_n)$ is a closed, convex and solid neighbourhood of 0 in E_α. Clearly, $i_\alpha^{-1}(B)$ is solid in E. Hence $i^{-1}(B)$ is a neighbourhood of 0 in E which implies that B is a neighbourhood of 0 in E. This shows that (E,C,u) is a C.O.Q. vector lattice.

Corollary 8. The locally convex direct sum of a sequence of order-(DF)-vector lattices is an order-(DF)-vector lattice.

Proof: Let $\{(E_n, C_n, u_n):\ n \geq 1\}$ be a sequence of order-(DF)-vector lattices. Let $E = \oplus_n E_n$, $C = \oplus_n C_n$ and $u = \oplus_n u_n$. Then (E,C,u) is a C.O.Q. vector lattice by Theorem 3. Let $\{B_i^{(n)};\ i \geq 1\}$ be a fundamental sequence of order-bounded sets in (E_n, C_n).

Then the family $\{\bigoplus_{n=1}^{k} B_{i_n}^{(n)}\}_{k\geq 1}$ is a fundamental sequence of order-bounded sets in (E,C) and hence (E,C,u) is an order-(DF)-vector lattice.

Proposition 13. Let (E,C,u) be a C.O.Q. vector lattice and (F,K,v) any l.c. vector lattice. Let f be a positive, linear, continuous and almost open map of E into F. Then (F,K,v) is a C.O.Q. vector lattice.

Proof: Let $B = \bigcap_{n\geq 1} V_n$ be an order-bornivorous N-barrel in (F,K,v). Clearly, $f^{-1}(B) = \bigcap_{n\geq 1} f^{-1}(V_n)$. Since f is positive and linear, it follows that $f^{-1}(B)$ is order-bornivorous. Since f is continuous, it follows that $f^{-1}(B)$ is an order-bornivorous N-barrel in (E,C,u) and hence a neighbourhood of 0. Now, as in the proof of Proposition 9, Chapter 5, we conclude that B is a neighbourhood of 0 in F.

A particular case of Proposition 13 is the following:

Corollary 9. Let (E,C,u) be a C.O.Q. vector lattice and M a closed lattice-ideal in E. Then E/M is a C.O.Q. vector lattice.

Corollary 10. Let (E,C,u) be an order-(DF)-vector lattice and M a closed lattice-ideal in E. Then E/M is an order-(DF)-vector lattice.

Proof: In view of Corollary 9, E/M is a C.O.Q. vector lattice. If $\{B_n;\ n \geq 1\}$ is a fundamental sequence of order-bounded sets in E, then it follows that $\{g(B_n);\ n \geq 1\}$, where g is the canonical map of E onto E/M, is a fundamental sequence of order-bounded sets in E/M. Hence E/M is an order-(DF)-vector lattice.

Proposition 14. The product of an arbitrary family of C.O.Q. vector lattices is again a C.O.Q. vector lattice.

Proof: Let $(E,u) = \prod_{\alpha\in\Phi} (E_\alpha,u_\alpha)$ be the product of a family $\{(E_\alpha,u_\alpha)\}$ of C.O.Q. vector lattices. If v_α is the finest locally solid topology on E_α, then (E_α,v_α) is clearly an order-quasibarrelled vector lattice. Hence $(E,v) = \prod_{\alpha\in\Phi}(E_\alpha,v_\alpha)$ is order-quasibarrelled. Let $B = \bigcap_{n\geq 1} V_n$ be a solid N-barrel in (E,u). Then it is also a solid N-barrel in (E,v) and so a neighbourhood of 0 in (E,v). Thus, for some finite subset Φ_0 of Φ, $\prod_{\alpha\in\phi\setminus\phi_o} E_\alpha \subset B$. Since $\prod_{\alpha\in\Phi_0} (E_\alpha,u_\alpha)$ is a C.O.Q. vector lattice, we conclude that $\prod_{\alpha\in\Phi} (E_\alpha,u_\alpha)$ is a C.O.Q. vector lattice.

The following example shows that a lattice ideal of a C.O.Q. vector lattice need not be of the same sort.

Example 8. Consider the Banach lattice E = C(I) and the lattice ideal F of C(I) as in (Chapter 4, Example 3). Clearly E is a C.O.Q. vector lattice. But F is not a C.O.Q. vector lattice under the induced topology; for if it were, then it would be an order-quasibarrelled vector lattice by Corollary 5, which is not true (Chapter 4, Example 3).

Remark 2. The above example also shows that a lattice ideal of an order-(DF)-vector lattice need not be an order-(DF)-vector lattice under the induced topology. Furthermore, F is another example of a countably quasibarrelled l.c. vector lattice which is not a C.O.Q. vector lattice.

Proposition 15. Let (E,C,u) be a C.O.Q. vector lattice such that C' is a strict $\mathcal{B}$-cone in (E',σ(E',E)). If M is a lattice ideal of countable codimension in E, then M is a C.O.Q. vector lattice under the induced topology.

Proof: Clearly, M is an l.c. vector lattice under the induced topology. By Proposition 3, (E,C,u) is a countably barrelled l.c. vector lattice and so M itself is countably barrelled (Chapter 5, Proposition 11). This implies that M is a C.O.Q. vector lattice under the induced topology.

Corollary 11. Let (E,C,u) be an order-(DF)-vector lattice such that C' is a strict $\mathcal{B}$-cone in (E',σ(E',E)). If M is a lattice ideal of countable codimension in E, then M is an order-(DF)-vector lattice under the induced topology.

Proof: By Proposition 14, (E,C,u) is a C.O.Q. vector lattice. If $\{B_n; n \geq 1\}$ is a fundamental sequence of order-bounded sets in E, then $\{M \cap B_n; n \geq 1\}$ is a fundamental sequence of order-bounded sets in M. This proves the Corollary.

Remark 3. All the results of this section, obviously, remain valid for order-(DF)-vector spaces (respectively, order-(DF)-

vector lattices).

§4. C.O.Q. vector spaces and Analogue of the Banach-Steinhaus Theorem.

Definition 5. An ordered l.c. vector space (E,C,u) is called a C.O.Q. vector space if every order-bornivorous N-barrel in E is a neighbourhood of 0.

Clearly every order-quasibarrelled vector space is a C.O.Q. vector space. The following theorem shows that the converse is true under a certain condition.

Theorem 4. A separable C.O.Q. vector space is an order-quasibarrelled vector space.

Proof: Let (E,C,u) be a separable C.O.Q. vector space and B an order-bornivorous barrel in E. Then, following the same argument as that in the proof of Proposition 5, Chapter 5, we can show that B is a neighbourhood of 0 in E.

Corollary 12. A separable order-(DF)-vector space is an order-quasibarrelled vector space.

Corollary 13. A separable C.O.Q. vector lattice (in particular, order-(DF)-vector lattice) is an order-quasibarrelled vector lattice.

Proposition 16. Let (E,C,u) be an ordered l.c. space with generating cone C, and let $E' \subset E^b$. Then the following statements are equivalent:

(a) (E,C,u) is a C.O.Q. vector space.

(b) Each $0(E',E)$-bounded subset of E' which is the countable union of equicontinuous subsets of E' is itself equicontinuous.

Proof: We observe that a subset H of E' is $0(E',E)$-bounded iff H° is order-bornivorous; therefore the equivalence of (a) and (b) follows as in (Chapter 5, Theorem 1).

We now obtain the analogue of the Banach-Steinhaus theorem for C.O.Q. vector spaces (C.O.Q. vector lattices).

Theorem 5. Let (E,C,u) be a C.O.Q. vector space (C.O.Q. vector lattice) and (F,K,v) any ordered l.c. space with normal cone (respectively, any l.c. vector lattice). Let $\{H_n\}$ be a countable family of equicontinuous sets of positive linear maps (respectively, lattice homomorphisms) of E into F such that $H = \bigcup_{n\geq 1} H_n$ is pointwise bounded. Then H is equicontinuous.

Proof: (a) Let (E,C,u) be a C.O.Q. vector space and (F,K,v) any ordered l.c. space with normal cone. Let V be a closed, circled, convex and full neighbourhood of 0 in F, and $W_n = \bigcap_{f\in H_n} f^{-1}(V)$. Then it easily follows that $W = \bigcap_{n\geq 1} W_n$ is an N-barrel in E. To show that W is order-bornivorous, let [x,y] be any order interval in E. Since H is pointwise bounded, there exists $\lambda > 0$ such that $f(x), f(y) \in \lambda V$ for all $f \in H$. But then $[f(x),f(y)] \subset \lambda V$ for all $f \in H$, because V is full. Since f is positive and linear, it follows that $f([x,y]) \subset [f(x),f(y)] \subset \lambda V$

for all $f \in H$. From this, it follows that $W = \bigcap_{n \geq 1} W_n = \bigcap_{f \in H} f^{-1}(V)$ is order-bornivorous. Hence W is a neighbourhood of 0 in E which in turn implies that H is equicontinuous.

(b) Let (E,C,u) be a C.O.Q. vector lattice and (F,K,v) any l.c. vector lattice. Let V be a closed, convex and solid neighbourhood of 0 in F, $W_n = \bigcap_{f \in H_n} f^{-1}(V)$, and $W = \bigcap_{n \geq 1} W_n = \bigcap_{f \in H} f^{-1}(V)$. Since each $f \in H$ is a continuous lattice homomorphism, it easily follows that W is a solid N-barrel in E and hence a neighbourhood of 0. This implies that H is equicontinuous.

As an immediate consequence, we have the following:

<u>Corollary 14</u>. Let (E,C,u) and (F,K,v) be as in Theorem 5. If $\{f_n\}$ is a pointwise bounded sequence of continuous linear maps (continuous lattice homomorphisms) of E into F, then $\{f_n\}$ is equicontinuous.

<u>Corollary 15</u>. Let (E,C,u) and (F,K,v) be as in Theorem 5. Suppose that the cone K in F is closed. If $\{f_n\}$ is a sequence of continuous positive linear maps (continuous lattice homomorphisms) of E into F such that $\{f_n\}$ converges pointwise to a map f of E into F, then f is a continuous positive linear map (continuous lattice homomorphism).

<u>Proof</u>: Clearly f is linear (respectively, lattice homomorphism). Since the cone K in F is closed, it follows that f is positive. By Corollary 14, $\{f_n\}$ is equicontinuous and hence f is continuous (Chapter 1, Corollary 6).

§5. Tensor Products.

Let E_1 and E_2 be l.c. spaces ordered by generating cones K_1 and K_2 respectively. Suppose that K_p is the projective cone in $E_1 \otimes E_2$. If S and T are order-bounded subsets of E_1 and E_2 respectively, then the convex circled hull $\Gamma(S \otimes T)$ of $S \otimes T$ is order-bounded for the order structure determined by K_p. The partial mappings g_x and g_y, $x \in E_1$, $y \in E_2$, associated with the canonical bilinear mapping g of $E_1 \times E_2$ into $E_1 \otimes E_2$ are order-bounded.

<u>Theorem 6</u>. Let (E_1,K_1,u_1) and (E_2,K_2,u_2) be C.O.Q. vector spaces, with generating cones. Let K_p be the projective cone in $E_1 \otimes E_2$. Then $(E_1 \otimes E_2,K_p,u_i)$, where u_i is the inductive (tensor product) topology, is a C.O.Q. vector space.

<u>Proof</u>: Let $B = \bigcap_{n\geq 1} V_n$ be an order-bornivorous N-barrel in $E_1 \otimes E_2$. Let g_x and g_y $(x \in E_1, y \in E_2)$ be the partial mappings associated with the canonical bilinear map g: $E_1 \times E_2 \to E_1 \otimes E_2$. Clearly $g_x^{-1}(B) = \bigcap_{n\geq 1} g_x^{-1}(V_n)$. Since g_x $(x \in E_1)$ is continuous and order-bounded, it follows that $g_x^{-1}(B)$ is an order-bornivorous N-barrel in E_2. Hence $g_x^{-1}(B)$ is a neighbourhood of 0 in E_2. Similarly $g_y^{-1}(B)$, $y \in E_2$, is a neighbourhood of 0 in E_1. This implies that B is an u_i-neighbourhood of 0 in $E_1 \otimes E_2$. Thus $E_1 \otimes E_2$ is a C.O.Q. vector lattice.

<u>Corollary 16</u>. Let (E_1,K_1,u_1) and (E_2,K_2,u_2) be order-quasi barrelled vector spaces with generating cones. Suppose that

K_p is the projective cone in $E_1 \otimes E_2$. Then $(E_1 \otimes E_2, K_p, u_i)$ is an order-quasibarrelled vector space.

Suppose that (E_1, u_1) and (E_2, u_2) are l.c. spaces ordered by generating cones K_1 and K_2 respectively and that the closure $\overline{K}_p$ of K_p in $E_1 \otimes E_2$ is a cone. If S and T are order-bounded subsets of E_1 and E_2 respectively, then the closure $\overline{\Gamma}(S \otimes T)$ of the convex circled hull of $(S \otimes T)$ in $E_1 \otimes_p E_2$ is order-bounded for the order structure determined by $\overline{K}_p$.

If P_1 and P_2 are the families of all order-bounded subsets of E_1 and E_2 respectively, then the $P_1 \times P_2$-topology on $\mathfrak{B}(E,F)$, also called the topology of bi-order-bounded convergence, is coarser than the topology $0(\mathfrak{B}(E_1,E_2), E_1 \otimes E_2)$. Naturally the following question [42] arises:

When does the topology of bi-order-bounded convergence on $\mathfrak{B}(E_1,E_2)$ coincide with the topology $0(\mathfrak{B}(E_1,E_2), E_1 \otimes E_2)$ on $\mathfrak{B}(E_1,E_2)$? Equivalently, when is each order-bounded set in $E_1 \otimes E_2$ for $\overline{K}_p$ contained in a set of the form $\overline{\Gamma}(S \otimes T)$ for suitable order-bounded sets S and T in E_1 and E_2 respectively?

Here we give an affirmative answer to this question when E_1 and E_2 are order-(DF)-vector spaces.

<u>Theorem 7.</u> Let (E_1, K_1, u_1) and (E_2, K_2, u_2) be as above, and suppose $\mathfrak{B}(E_1,E_2) \subseteq B^b(E_1,E_2)$. If (E_1, K_1, u_1) and (E_2, K_2, u_2) are order-(DF)-vector spaces, then the topology of bi-order-bounded convergence on $\mathfrak{B}(E_1,E_2)$ coincides with $0(\mathfrak{B}(E_1,E_2), E_1 \otimes E_2)$.

Proof: Clearly $E_1 \times E_2$ is an order-(DF)-vector space with a generating cone. Since $E_1 \times E_2$ has a fundamental sequence of order-bounded sets, it follows that the topology u of bi-order-bounded convergence is metrizable. Since u is coarser than $O(\mathcal{B}(E_1,E_2),E_1 \otimes E_2)$, it is enough if we show that the identity mapping of $(\mathcal{B}(E_1,E_2),u)$ onto $(\mathcal{B}(E_1,E_2),O(\mathcal{B}(E_1,E_2),E_1 \otimes E_2))$ is continuous. For this, in turn, it is enough to show that every u-null sequence $\{f_n\}$ in $\mathcal{B}(E_1,E_2)$ is $O(\mathcal{B}(E_1,E_2),E_1 \otimes E_2)$-bounded. Consider $V = \bigcap_{n \geq 1} f_n^{-1}([-\varepsilon,\varepsilon])$, $\varepsilon > 0$. Clearly V is an N-barrel in $E_1 \times E_2$. Since $\{f_n\}$ is u-bounded in $\mathcal{B}(E_1,E_2)$, it follows that V is order-bornivorous. Hence V is a neighbourhood of 0 in $E_1 \times E_2$. This implies that $\{f_n\}$ is equicontinuous on $E_1 \times E_2$ and hence equicontinuous in the dual $\mathcal{B}(E_1,E_2)$ of $E_1 \otimes E_2$. Thus $\{f_n\}$ is $O(\mathcal{B}(E_1,E_2),E_1 \otimes E_2)$-bounded. This completes the proof.

§6. Open problem.

Problem 1. Is the assumption "C' is a strict $\mathcal{B}$-cone in $(E',\sigma(E',E))$" in Proposition 15 superfluous?

CHAPTER IX

ORDER-QUASIULTRABARRELLED VECTOR LATTICES AND SPACES

The concepts of barrelled and quasibarrelled l.c. spaces led W. Roberton [44] and Iyahan [19] respectively to introduce the concepts of ultrabarrelled and quasiultrabarrelled spaces in situations where local convexity is not assumed. In this chapter, we introduce and study the concept of order-quasi-ultrabarrelled vector lattice (vector space) which replaces that of order quasibarrelled vector lattices (vector spaces) in non-locally convex situations. Most of the results of this chapter are based on the paper [16].

§1. O.Q.U. vector lattices.

Definition 1. Let (E,C,u) be an ordered t.v. space (a t.v. lattice). A closed circled (respectively closed solid) subset B of E is said to be an order-bornivorous ultrabarrel (respectively, solid ultrabarrel) if there exists a sequence $\{B_n\}$ of closed, circled, order-bornivorous (closed, solid, absorbing) subsets of E such that $B_1 + B_1 \subseteq B$ and $B_{n+1} + B_{n+1} \subseteq B_n$ for all $n \geq 1$.

The sequence $\{B_n\}$ is called a defining sequence for B.

Definition 2. A t.v. lattice (E,C,u) is called an order-quasiultrabarrelled vector lattice (abbreviated to O.Q.U. vector lattice) if each order-bornivorous ultrabarrel in E is a neighbourhood of 0.

Theorem 1. A t.v. lattice (E,C,u) is an O.Q.U. vector lattice iff each solid ultrabarrel in E is a neighbourhood of 0.

Proof: Assume that (E,C,u) is an O.Q.U. vector lattice. Let B be a solid ultrabarrel in (E,C,u) with $\{B_n\}$ as a defining sequence. Since a solid set is circled and a solid absorbing set is order-bornivorous, it follows that B is an order-bornivorous ultrabarrel in (E,C,u) and hence a neighbourhood of 0. Conversely, assume that each solid ultrabarrel in E is a neighbourhood of 0. Let V be an order-bornivorous ultrabarrel with $\{V_n\}$ as a defining sequence. We show as has been shown in (Chapter 4, Theorem 1) that K(V) and $K(V_n)$ are closed. Since V_n is order-bornivorous it follows that $K(V_n)$ is order-bornivorous (Chapter 4, Lemma 1). We now show that

$$K(V_1) + K(V_1) \subseteq K(V) \quad \text{and}$$

$$K(V_{n+1}) + K(V_{n+1}) \subset K(V_n) \quad \text{for all } n \geq 1.$$

Let $x,y \in K(V_1)$. Then $[-|x|,|x|] + [-|y|,|y|] \subset V_1 + V_1 \subset V$. Since E has the decomposition property, we get

$$[-(|x| + |y|),(|x| + |y|)] \subset V.$$

But then

$$[-|x + y|,|x + y|] \subset V.$$

Hence

$$x+y \in K(V), \quad \text{i.e. } K(V_1) + K(V_1) \subset K(V).$$

Similarly, we have

$$K(V_{n+1}) + K(V_{n+1}) \subseteq K(V_n)$$

for all n. Thus $K(V_n)$ is a solid ultrabarrel with $\{K(V_n)\}$ as a defining sequence. Hence K(V) is a neighbourhood of 0 in E. But $K(V) \subset V$ and so V is a neighbourhood of 0 in E. This completes the proof.

Theorem 2. Let (E,C,u) be a t.v. lattice. The following statements are equivalent:

(a) (E,C,u) is an O.Q.U. vector lattice.

(b) Any locally solid linear topology on E with a base of u-closed neighbourhoods of 0 is coarser than u.

(c) Each lower-semi-continuous lattice $\mathcal{F}$-seminorm on E is continuous.

Proof: (a)$\Leftrightarrow$(b). This follows immediately from the definitions, because a t.v. lattice (E,C,u) is an O.Q.U. vector lattice iff each locally solid linear topology with a base of neighbourhoods of 0 consisting of u-ultrabarrels is coarser than u.

(b)$\Rightarrow$(c). Let p be a lower semi-continuous $\mathcal{F}$-seminorm on E. p determines a topology v on E with a base of u-closed solid neighbourhoods of 0. By (b), v is coarser than u and so p is continuous.

(c)$\Rightarrow$(a). Let B_0 be a solid ultrabarrel with $\{B_n\}$ as a defining sequence in E. Let α be a dyadic rational number. Define U_α as in (Chapter 3, Theorem 2) and $\overline{U}_\alpha = cl_u U_\alpha$. Now define

$$p(x) = \inf\{\alpha > 0: \ x \in \overline{U}_\alpha\}.$$

Clearly p is a lower semi-continuous lattice $\mathfrak{F}$-seminorm and hence, by (c), it is continuous and therefore B_0 is a neighbourhood of 0 in E. This completes the proof.

Proposition 1. (a) Every ultrabarrelled t.v. lattice is an O.Q.U. vector lattice; (b) Every O.Q.U. vector lattice is a quasiultrabarrelled t.v. lattice.

Proof: (a) This is obvious in view of the definitions concerned.

(b) This follows immediately if we note that the positive cone in a t.v. lattice is normal and so every bornivorous ultrabarrel is, a fortiori, an order-bornivorous ultrabarrel.

Every complete metrizable t.v. lattice is an ultrabarrelled t.v. lattice (Chapter 3, Corollary 1) and hence, by Proposition 1, part (a), it is an O.Q.U. vector lattice. Examples will be given later to show that the converse in each of the statements (a) and (b) of Proposition 1 is false.

If u is a locally solid linear topology on a vector lattice (E,C) then we use u^∞ to denote the locally convex topology derived from u, i.e. the finest locally convex topology coarser than u.

Proposition 2. If (E,C,u) is an O.Q.U. vector lattice, then (E,C,u^{∞}) is an order-quasibarrelled vector lattice.

Proof: Let B be an order-bornivorous barrel in (E,C,u^{∞}). Then B is an order-bornivorous ultrabarrel in (E,C,u) and hence a u^{∞}-neighbourhood of 0. The set B must then be a u^{∞}-neighbourhood of 0 and so (E,C,u^{∞}) is an order-quasibarrelled vector lattice.

Corollary 1. Any locally convex O.Q.U. vector lattice is an order quasibarrelled vector lattice.

An example will be given later to show that the converse in Corollary 1 is false.

Proposition 3. Every order-unit-normed vector lattice is an O.Q.U. vector lattice.

Proof: Let (E,C,u) be an order-unit normed vector lattice with order-unit e. Then $\{n[-e,e];\ n \geq 1\}$ is a fundamental system of order-bounded sets in E. Let B be an order-bornivorous ultrabarrel in E. Then there exists $\lambda > 0$ such that $\lambda[-e,e] \subseteq B$. But the scalar multiplies of [-e,e] form a neighbourhood basis of 0 for u. Hence B is a neighbourhood of 0, and this completes the proof.

The following example shows that an O.Q.U. vector lattice need not be ultrabarrelled.

Example 1. Consider the Banach lattice m and the vector subspace E of m as in (Chapter 4, Example 1). As has been shown there, E is not barrelled and hence it is not ultrabarrelled (Chapter 3,

Corollary 10). But E is an order unit-normed vector lattice and hence, by Proposition 3, it is an O.Q.U. vector lattice.

We now give an example of a quasiultrabarrelled t.v. lattice which is not an O.Q.U. vector lattice.

Example 2. Consider the normed lattice ϕ of all real sequences with only finitely many non-zero components. As has been shown (Chapter 4, Example 2), ϕ is not order-quasibarrelled and hence by Proposition 2, it is not an O.Q.U. vector lattice either. However ϕ being a normed vector lattice is a quasi-ultrabarrelled t.v. lattice. (Chapter 3, Remark 7 and Corollary 12).

It is clear from (Chapter 3, Proposition 4) that a (Hausdorff) sequentially complete almost convex quasiultra-barrelled t.v. lattice is an ultrabarrelled t.v. lattice and hence, by Proposition 1, part (b), every (Hausdorff) sequentially complete almost convex O.Q.U. vector lattice is ultrabarrelled. It is interesting to find conditions under which a quasi-ultrabarrelled t.v. lattice is an O.Q.U. vector lattice, but not ultrabarrelled.

§2. Permanence Properties.

Lemma 1. Let $\{(E_n, C_n): n \geq 1\}$ be a countable family of ordered vector spaces, u_n a linear topology on E_n for each n such that C_n is normal with respect to u_n. Then $C = \oplus_n C_n$ is normal with respect to the *-direct sum topology $u = *\text{-}\oplus_n u_n$.

<u>Proof</u>: Let $\eta_n = \{V^{(n)}\}$ be a basis of circled neighbourhoods of 0 in E_n such that $0 \le x^{(n)} \le y^{(n)}$, $y^{(n)} \varepsilon V^{(n)}$ implies $x^{(n)} \varepsilon V^{(n)}$. Then it can easily be shown that $0 \le x \le y$, $y \varepsilon W = \bigcup_J (\sum_{n \varepsilon J} V^{(n)})$, J a finite subset of $\mathbb{N}$, implies that $x \varepsilon W$. But as $V^{(n)}$ runs through η_n, the sets W form a neighbourhood basis of 0 for (E,C,u) (Chapter 1, Proposition 6) and hence C is normal for u.

The proof of the following lemma is similar to that of Lemma 7 of Chapter 4 and hence is omitted.

<u>Lemma 2</u>. Let u be a linear topology on a vector lattice (E,C) such that the positive cone C in E is normal with respect to u. Let $\eta = \{U\}$ be a neighbourhood basis of 0 in E for u, consisting of circled full sets, and let $\eta_s = \{K(U); U \varepsilon \eta\}$. Then there exists a unique locally solid linear topology u_s on E such that η_s is a neighbourhood basis of 0 in E for u_s. Furthermore, u_s is finer than u, and u_s is the greatest lower bound of all locally solid linear topologies which are finer than u.

The above lemma leads us to the following:

<u>Definition 3</u>. u_s, as defined above, is called the locally solid linear topology on E associated with u.

The proofs of the following proposition and its corollary, being similar to those of Lemma 10 and Corollary 8 of Chapter 4 are omitted. The only difference is that we do not have convexity here and so we have to use Lemma 1 instead of Proposition 36 of Chapter 1.

Proposition 4. Let $\{(E_n,C_n);\ n \in \mathbb{N}\}$ and u_n be as in Lemma 2. Let $u_{n,s}$ denote the locally solid linear topology on E_n associated with u_n for each $n \in \mathbb{N}$ and let u'_s be the locally solid linear topology on $E = \oplus_n E_n$ associated with $u = *\text{-}\oplus_n u_n$. Then u'_s is coarser than $*\text{-}\oplus_n u_{n,s}$.

Corollary 2. Let u_n be the locally solid linear topology on E_n for each $n \in \mathbb{N}$. Then the *-direct sum topology $u = *\text{-}\oplus_n u_n$ on E is a locally solid linear topology and hence (E,C,u) is a t.v. lattice.

Theorem 3. The *-direct sum of a countable family of O.Q.U. vector lattices is an O.Q.U. vector lattice.

Proof: Let $\{(E_n,C_n,u_n): n \geq 1\}$ be a countable family of O.Q.U. vector lattices, $E = \oplus_n E_n$, $C = \oplus_n C_n$ and $u = *\text{-}\oplus_n u_n$. By Corollary 2, (E,C,u) is a t.v. lattice. Let $i_n: E_n \to E$ be the injection map for each n. Let V be a solid ultrabarrel in (E,C,u). Since each i_n is a continuous lattice homomorphism, it follows that $i_n^{-1}(V)$ is a solid ultrabarrel in E_n and hence a neighbourhood of 0 in E. This implies that V is a neighbourhood of 0 in (E,C,u) and so (E,C,u) is an O.Q.U. vector lattice.

Proposition 5. Let (E,C,u) and (F,K,v) be t.v. lattices and f a lattice homomorphism of E into F.

(a) If (E,C,u) is an O.Q.U. vector lattice, then f is almost continuous.

(b) If (F,K,v) is an O.Q.U. vector lattice and f onto, then f is almost open.

Proof: (a) Let V be a closed solid neighbourhood of 0 in F. Since f is a lattice homomorphism, it follows that $f^{-1}(V)$ and hence $\overline{f^{-1}(V)}$ is a solid set in E. Furthermore, $\overline{f^{-1}(V)}$, being an ultrabarrel (Chapter 3, Proposition 1) is a neighbourhood of 0 in E and so f is almost continuous.

(b) Let W be a closed solid neighbourhood of 0 in E. Since f is a lattice homomorphism, it follows that f(W) and so $\overline{f(W)}$ is a solid subset of F. Thus by the same argument as in (a) $\overline{f(W)}$ is a neighbourhood of 0 in F and so f is almost open.

Proposition 6. Let (E,C,u) be an O.Q.U. vector lattice and (F,K,v) any t.v. lattice. If f is a positive, linear, continuous and almost open mapping of E into F, then (F,K,v) is an O.Q.U. vector lattice.

Proof: Let V be an order-bornivorous ultrabarrel in F. Since f is positive and linear, it easily follows that $f^{-1}(V)$ is an order-bornivorous set in E. Furthermore, since f is continuous $f^{-1}(V)$ being an ultrabarrel in E (Chapter 3, Proposition 1) is a neighbourhood of 0 in E. Since f is almost open and $\overline{f(f^{-1}(V))} \subset V$, we conclude that V is a neighbourhood of 0 in F.

Corollary 3. Let (E,C,u) be an O.Q.U. vector lattice and M a closed lattice ideal in E. Then E/M is an O.Q.U. vector lattice.

The following example shows that a lattice ideal of an O.Q.U. vector lattice need not be an O.Q.U. vector lattice.

Example 3. Consider the Banach lattice E = C(I) and the subspace F of E as in (Chapter 4, Example 3). Clearly E being ultrabarrelled is an O.Q.U. vector lattice, and F is a lattice ideal in E. But F is not order-quasibarrelled (Chapter 4, Example 3) and hence, by Corollary 1, it is not an O.Q.U. vector lattice.

Adopting the techniques of (Chapter 3, Proposition 4') we can show that the product of any family of O.Q.U. vector lattices is an O.Q.U. vector lattice.

§3. Analogue of the Banach-Steinhaus theorem.

Definition 3. An ordered t.v. space (E,C,u) is said to be an O.Q.U. vector space if every order-bornivorous ultrabarrel is a neighbourhood of 0 in E.

Theorem 4. Let (E,C,u) be an O.Q.U. vector space (O.Q.U. vector lattice) and (F,K,v) any ordered t.v. space with normal cone (respectively t.v. lattice). If H is a pointwise bounded set of continuous positive linear maps (respectively, continuous lattice homomorphism) of E into F, then H is equicontinuous.

Proof: (a) Assume that (E,C,u) is an O.Q.U. vector space. Let V be a closed, circled, full neighbourhood of 0 in F. Then $W = \bigcap_{f \in H} f^{-1}(V)$ is an ultrabarrel. Furthermore, since H is pointwise bounded it easily follows that W is order-bornivorous. But then W is a neighbourhood of 0 in E, and hence H is equicontinuous.

(b) Assume that (E,C,u) is an O.Q.U. vector lattice. Let V be a closed solid neighbourhood of 0 in F. Then $W = \bigcap_{f \in H} f^{-1}(V)$ is a solid subset of E, because each $f \in H$ is a lattice homomorphism. Clearly W is an ultrabarrel in E. Hence, by Theorem 1, W is a neighbourhood of 0 in E and so H is equicontinuous.

Corollary 4. Let (E,C,u) and (F,K,v) be as in Theorem 4. Let $\{f_n; n \geq 1\}$ be a pointwise bounded sequence of continuous positive linear maps (continuous lattice homomorphisms) of E into F. Then $\{f_n\}$ is equicontinuous.

Corollary 5. Let (E,C,u) and (F,K,v) be as in Theorem 4. Suppose that the cone K in F is closed. If $\{f_n; n \geq 1\}$ is a sequence of continuous positive linear maps (continuous lattice homomorphisms) of E into F such that it converges pointwise to a map $f: E \to F$, then f is continuous, positive and linear.

The isomorphism theorem (Chapter 4, Theorem 8) remains

valid for an O.Q.U. vector space as well by Corollary 5. Furthermore, using Corollary 4, we can easily establish the following:

Proposition 5. Let $\{x_i, f_i\}$ be a positive Schauder basis in an O.Q.U. vector space (E,C,u). Let $\{T_n\}$ be a sequence of maps $T_n: E \to E$ defined by

$$T_n(x) = \sum_{i=1}^{n} f_i(x)x_i, \quad x \in E, \ n \geq 1$$

Then $\{T_n\}$ is equicontinuous on compact sets and converges uniformly to the identity map.

We have remarked in Chapter 3 that if the conclusion of Theorem 5, in Chapter 3, is true, then E is ultrabarrelled. The technique used by Waelbroeck [54], to prove it, is now adopted to obtain the following analogous result.

Theorem 5. Let (E,C,u) and (F,K,v) be any t.v. lattices such that every pointwise bounded set of continuous lattice homomorphisms of E into F is equicontinuous. Then (E,C,u) is an O.Q.U. vector lattice.

Proof: Let B_0 be a solid ultrabarrel with $\{B_n\}$ as a defining sequence in E. We now construct a metrizable t.v. lattice (F,K,v) with $\{V_n;\ n \geq 0\}$ as a base of solid neighbourhoods of 0, and a pointwise bounded family of continuous mappings $f_\lambda: E \to F$ in such a way that $B_n = \bigcap_{\lambda \in \Lambda} f_\lambda^{-1}(V_n)$ where Λ is the set of all sequences $\{W_{n\lambda} : n \geq 1\}$, $\lambda \in \Lambda$, of solid neighbourhoods of 0 such that $W_{n+1,\lambda} + W_{n+1,\lambda} \subset W_{n,\lambda}$ for all n. For each $\lambda \in \Lambda$, let $M_\lambda = \bigcap_{n \geq 1} (W_{n\lambda} + B_n)$. We observe that M_λ is a lattice ideal of E and $W_{n,\lambda} + B_n$ is a union of cosets of M_λ. Let $E/M_\lambda = F_\lambda$ and $V_{n,\lambda}$ be the quotient image of $W_{n,\lambda} + B_n$ in F_λ. Further, let $F = \bigoplus_{\lambda \in \Lambda} F_\lambda$ and V_n the set of families $\{x_\lambda\}_{\lambda \in \Lambda}$ which belong to F and are such that each $x_\lambda \in V_{n,\lambda}$. Then $\{V_n\}_{n \geq 1}$ is a basis of a solid absorbing filter on F and $V_{n+1} + V_{n+1} \subseteq V_n$ so that it is the filter of neighbourhoods of 0 for some locally solid linear topology v on F. Now define the map $f_\lambda : E \to F$ by

$$\begin{array}{ccc} E & \xrightarrow{\;f_\lambda\;} & F \\ & {\scriptstyle\phi}\searrow \quad \nearrow{\scriptstyle i_\lambda} & \\ & F_\lambda & \end{array}$$

We observe that $f_\lambda^{-1}(V_n) = W_{n\lambda} + B_n$. Since the sets $W_{n\lambda}$ are neighbourhoods of 0, it follows that each f_λ is continuous. Furthermore, since each set B_n is absorbing the family of mappings f_λ is pointwise bounded. Hence we conclude that

$$\bigcap_{\lambda \in \Lambda} f_\lambda^{-1}(V_n) = \bigcap_{\lambda \in \Lambda} (W_{n\lambda} + B_n) = B_n$$

is a neighbourhood of 0 in E. This completes the proof.

By using Corollary 4, we now give an example to show that the converse in Corollary 1 is false.

<u>Example 4</u>. The complete metrizable t.v. space $(\ell^{1/2}, u)$, as defined in (Chapter 3, Example 2) is in fact a t.v. lattice

with positive cone $K = \{x = \{x_i\} \in \ell^{\frac{1}{2}};\ x_i \geq 0$ for all $i\}$. As we have seen there, $(\ell^{\frac{1}{2}}, u^{oo})$ is barrelled and so $(\ell^{\frac{1}{2}}, K, u^{oo})$ is an order-quasibarrelled vector lattice. We show, however, that it is not an O.Q.U. vector lattice. For $x = \{x_i\} \in \ell^{\frac{1}{2}}$, define $f_n(x) = (x_1, x_2, \ldots, x_n, 0, 0, \ldots)$. Then each f_n is a continuous lattice homomorphism of $(\ell^{\frac{1}{2}}, K, u^{oo})$ into $(\ell^{\frac{1}{2}}, K, u)$, and for each $x \in \ell^{\frac{1}{2}}$, $f_n(x) \to x$ under u. But, since the identity map of $(\ell^{\frac{1}{2}}, K, u^{oo})$ onto $(\ell^{\frac{1}{2}}, K, u)$ is not continuous, $\{f_n\}$ is not equicontinuous. Hence, in view of Corollary 4, we conclude that $(\ell^{\frac{1}{2}}, K, u^{oo})$ is not an O.Q.U. vector lattice.

§4. A Closed Graph Theorem.

Definition 4. Let (E,C,u) be an ordered t.v. space and (F,v) any t.v. space. A map $f: E \to F$ is called semibounded if f maps order-bounded subsets of E into v-bounded subsets of F.

Proposition 6. Let (E,C,u) be an O.Q.U. vector space and (F,v) any t.v. space. If $f: E \to F$ is a linear semi-bounded mapping, then it is almost continuous.

Proof: Let V be a circled neighbourhood of 0 in F and $[x,y]$ any order interval in E. Since f is semi-bounded, there exists $\lambda > 0$ such that $f([x,y]) \subset \lambda V$ and so $f^{-1}(V)$ is order-bornivorous in E. Then, surely, $\overline{f^{-1}(V)}$ is order-bornivorous in E and also an ultrabarrel in E. Hence $\overline{f^{-1}(V)}$ is a neighbourhood of 0 in E and so f is almost continuous.

Corollary 6. Let (E,C,u) be an O.Q.U. vector space and (F,K,v) any ordered t.v. space with normal cone K. If f: E → F is a positive linear mapping, then it is almost continuous.

Proof: Since f is positive and linear, it follows that f is order-bounded. But then f is semibounded since the cone K in F is normal and Proposition 6 applies.

Corollary 7. Let (E,C,u) be an O.Q.U. vector lattice and (F,K,v) any t.v. lattice. If f: E → F is a lattice homomorphism, then it is almost continuous.

Proof: Since a lattice homomorphism is positive and linear and the cone in a t.v. lattice is normal, Corollary 6 applies.

Theorem 6. Let (E,C,u) be an O.Q.U. vector space and (F,v) any complete metrizable t.v. space. If f: E → F is a linear semibounded mapping and has closed graph, then f is continuous.

Proof: In view of Proposition 6, f is almost continuous. Hence by (Chapter 1, Theorem 7), f is continuous.

Corollary 8. Let (E,C,u) be an O.Q.U. vector space and (F,K,v) a complete metrizable ordered t.v. space with normal cone K. If f: E → F is a positive linear map with closed graph, it is continuous.

Proof: f is almost continuous, by Corollary 6. Hence f is continuous by (Chapter 1, Theorem 7).

Let (E,C,u) be a t.v. lattice and B_0 a solid ultrabarrel

with $\{B_n\}_{n\geq 1}$ as a defining sequence in E. Then $\{B_n; n \geq 0\}$ is a base of neighbourhoods of 0 for a locally solid linear topology on E. If M is the lattice ideal $\bigcap_{n\geq 1} B_n$ of E and $\phi: E \to E/M$, the canonical map, it is easy to see that the locally solid topology v on E/M with 0-neighbourhood basis $\{\phi(B_n)\}$, is metrizable. Let $(\widetilde{E/M}, \widetilde{\phi(C)}, \tilde{v})$ denote the completion of $(E/M, \phi(C), v)$. We prove the following:

<u>Lemma 3</u>. The map ϕ of (E,C,u) into $(\widetilde{E/M}, \widetilde{\phi(C)}, \tilde{v})$ has closed graph.

<u>Proof</u>: Let G denote the graph of ϕ, and G_1 its closure in $(E,u) \times (\widetilde{E/M}, \tilde{v})$. Let $(x,y) \in G_1$. Since E/M is dense in $(\widetilde{E/M}, \tilde{v})$, for each n there exists some $z \in E$ such that

$$(*) \qquad \phi(z) \in y - \phi(x) + cl(\phi(B_{n+4})).$$

Since $(x,y) \in G_1$, for each u-neighbourhood U, we have

$$(x+U, y+cl(\phi(B_{n+4}))) \cap G \neq \emptyset.$$

Therefore, $y-\phi(x) \in \phi(U) + cl(\phi(B_{n+4}))$, and using this in (*) we find that

$$\phi(z) \in \phi(U) + cl(\phi(B_{n+3})).$$

Thus $\phi(z) \in \phi(U) + \phi(B_{n+3}) + \phi(B_{n+3})$, which is contained in $\phi(U) + \phi(B_{n+2})$. This implies that $z \in U + B_{n+2} + M$, and thus $z \in U + B_{n+1}$. Since this is true for all u-neighbourhoods U, and B_{n+1} is closed, it follows that $z \in B_{n+1}$. Now using this in (*), we find that $y - \phi(x) \in cl(\phi(B_n))$ and this completes the proof.

Theorem 7. Let (E,C,u) be a Hausdorff t.v. lattice and (F,K,v) any complete metrizable t.v. lattice. Then the following statements are equivalent:

(a) (E,C,u) is an O.Q.U. vector lattice.

(b) Each lattice homomorphism of E into F with closed graph is continuous.

Proof: (a)$\Rightarrow$(b): This follows from Corollary 7 and (Chapter 1, Theorem 7).

(b)$\Rightarrow$(a): Let B_0 be a solid ultrabarrel in (E,C,u) with $\{B_n; n \geq 1\}$ as a defining sequence. By Lemma 3, ϕ is continuous. Hence $\phi^{-1}(\phi(B_1))$ is a neighbourhood of 0 in (E,C,u). But $\phi^{-1}(\phi(B_1)) = B_1 + M \subset B$ and so B is a neighbourhood of 0. This implies that (E,C,u) is an O.Q.U. vector lattice.

BIBLIOGRAPHY

[1] Adasch, N.; Topologische Produkte gewisser topologischer Vektorräume; Math. Ann 189 (1970), 280-284.

[2] Arsove, M.G. and Edwards, R.E.; Generalized bases in topological linear spaces, Studia Math. 19 (1960), 95-113.

[3] Banach, S.; Théorie des opérations linéaires, Warsaw (1932).

[4] Bourbaki, N.; Espaces vectoriels topologiques, Chapters I-V Herman, Paris (1953,1955).

[5] Bozel, F. and Husain, T.; On isomorphisms of locally convex spaces with similar biorthogonal systems, Canadian Math. Bull. Vol. 16(2) (1973).

[6] De Wilde, M.; Sur les sous-espaces de codimension finie d'un espace linéaire a seminormes. Bull. Soc. Royale de Sci. Liege n° 9-10 (1969), 450-453.

[7] De Wilde, M. and Houet, C.; On increasing sequences of absolutely convex sets in locally convex spaces, Math. Ann. 192 (1971), 257-261.

[8] Dieudonné, J.; Sur les espaces de Köthe, J. Analyse Math. 1 (1951), 81-115.

[9] Edwards, R.E.; Functional analysis - Theory and applications, Holt, Rinehart and Winston, New York (1965).

[10] Grothendieck, A.; Sur les espaces (F) et (DF), Summa Brasil. Math, 3 (1954), 57-123.

[11] Horváth, J.; Topological vector spaces I, Addison-Wesley Publ. Co.; Reading, Mass. (1966).

[12] Husain, T.; The open mapping and closed graph theorems in topological vector spaces, Oxford Math. Monographs (1965).

[13] Husain, T.; Two new classes of locally convex spaces, Math. Ann. 166 (1966), 289-299.

[14] Husain, T. and Wong, Y.C.; On various types of barrelledness and the hereditary property of (DF)-spaces. Glasgow Math. Journal 17 (1976), 134-143.

[15] Husain, T. and Khaleelulla, S.M.; Countably order-quasibarrelled vector lattices and spaces, Mathematicae Japonicae 20 (1975), 3-15.

[16] Husain, T. and Khaleelulla, S.M.; Order-quasiultrabarrelled vector lattices and spaces, Periodica Mathematica Hungrica Vol. 6(4), (1975), 363-371.

[17] Husain, T. and Khaleelulla, S.M.; On countably, σ-, and sequentially barrelled spaces, Can. Math. Bull. 18 (1975), 431-432.

[18] Iyahen, S.O.; Some remarks on countably barrelled and countably quasibarrelled spaces, Proc. Edinburgh Math. Soc. (2), 15 (1966/67), 295-296.

[19] Iyahen, S.O.; On certain classes of linear topological spaces. Proc. London Math. Soc. (3), 18 (1968), 285-307.

[20] Iyahen, S.O.; On certain classes of linear topological spaces II, J. London Math. Soc. (2), 3 (1971), 609-617.

[21] Iyahen, S.O.; Corrigendum: On certain classes of linear topological spaces II, J. London Math. Soc. (2), 5 (1972), 740.

[22] Jameson, G.; Ordered linear spaces, Springer-Verlag; Berlin (1970).

[23] Johnson, W.B. and Dyer, J.; Isomorphisms generated by fundamental and total sets, Proc. Amer. Math. Soc. 22 (1969), 330-334.

[24] Jones, O.T. and Retherford, J.R.; On similar bases in barrelled spaces, Proc. Amer. Math. Soc. 18 (1967), 677-680.

[25] Kelley, J.L.; General topology, Van Nostrand; New York (1955).

[26] Khaleelulla, S.M.; Husain spaces, Tamkang J. Math. Vol 6, No. 2 (1975). 115-119.

[27] Klee, V.L.; Leray-Schauder theory without local convexity, Math. Ann. 141 (1960), 286-296.

[28] Klee, V.L.; Shrinkable neighbourhoods in Hausdorff linear spaces, Math. Ann. 141 (1960), 281-285.

[29] Kōmura, Y.; On linear topological spaces, Kumamoto J. Science, Series A, 5. No. 3 (1962), 148-157.

[30] Kōmura, Y.; Some examples on linear topological spaces, Math. Ann. 153 (1964), 150-162.

[31] Köthe, G.; Topological vector spaces I, Springer-Verlag; New York, Inc. (1969).

[32] Levin, M. and Saxon, S.; A note on the inheritance of properties of locally convex spaces by subspaces of countable codimension, Proc. Amer. Math. Soc. Vol.29, No. 1 (1971), 97-102.

[33] Mahowald, M. and Gould G.; Quasibarrelled locally convex spaces, Proc. Amer. Math. Soc. 2 (1960), 811-816.

[34] Marti, J.T.; Introduction to the theory of bases, Springer-Verlag, New York Inc. (1969).

[35] Marti, J.T.; On positive bases in order topological vector spaces, Archiv der Math. 22 (1971), 657-659.

[36] Marti, J.T.; and Sherbert, D.R. A note on bases in ordered locally convex spaces, Comment. Math. Helv, 45 (1970), 299-302.

[37] Morris, P.D. and Wulbert, D.E.; Functional representation of topological algebras, Pacific J. Math. Vol. 22, No. 2 (1967), 232-337.

[38] Nachbin, L.; Topological vector spaces of continuous functions, Proc. Nat. Acad. Sci. Wash. 40 (1954), 471-474.

[39] Namioka, I.; Partially ordered linear topological spaces, Memoirs of the Amer. Math. Soc., No. 24 (1957).

[40] Ng. K-F.; Solid sets in ordered topological vector spaces, Proc. London Math. Soc. (3), 22 (1971), 106-120.

[41] Peressini, A.L.; Ordered topological vector spaces, Harper and Row Publ.; New York (1967).

[42] Peressini, A.L. and Sherbert, D.R.; Ordered topological tensor products, Proc. London Math. Soc. (3), 19 (1969), 177-190.

[43] Robertson, A.P. and Robertson, W.J.; Topological vector spaces (second edition), Camb. tracts in Math., Camb. Univ. Press; England (1973).

[44] Robertson, W.J.; Completions of topological vector spaces, Proc. London Math. Soc (3), 8 (1958), 242-257.

[45] Saxon, S. And Levin M.; Every countable codimensional subspace of a barrelled space is barrelled, Proc. Amer. Math. Soc. Vol. 29, No. 1 (1971), 91-96.

[46] Saxon, S.A.; Nuclear and product spaces. Baire-like spaces and the strongest locally convex topology, Math. Ann. 197 (1972), 87-106.

[47] Schaefer, H.H.; Topological vector spaces, Springer-Verlag; New York-Heidelberg-Berlin(1971).

[48] Schmets, J.; Espaces C(X), évaluable, infra-évaluable and σ-évaluable, Bull. Soc. Roy. Sc. Liège, 40 (1971), 122-126.

[49] Shirota, J.; On locally convex vector spaces of continuous functions, Proc. Japan. Acad. 30 (1954), 294-299.

[50] Singer, I.; Bases in Banach spaces I, Springer-Verlag; New York-Heidelberg-Berlin (1970).

[51] Tweddle, I.; Some remarks on σ-barrelled spaces (unpublished).

[52] Valdivia, M.; Absolutely convex sets in barrelled spaces, Ann. Inst. Fourier, Grenoble 21, 2 (1971), 3-13.

[53] Valdivia, M.; On (DF)-spaces, Math. Ann. 43 (1971), 38-43.

[54] Waelbroeck, L.; Topological vector spaces and algebras. Lecture notes in Mathematics, Springer-Verlag; Berlin-Heidelberg-New York (1971).

[55] Warner, S.; The topology of compact convergence on continuous function spaces, Duke Math. J. 25 (1958), 265-282.

[56] Webb, J.H.; Sequential convergence in locally convex spaces, Proc. Camb. Phil. Soc. 64 (1968), 341-364.

[57] Webb, J.H.; Completeness and strong completeness in locally convex spaces, J. London Math. Soc. (2), 1 (1960), 767-768.

[58] Webb, J.H.; Countable codimensional subspaces of locally convex spaces, Proc. Edinburg Math. Soc. 18 (1972/73), 167-172.

[59] Wilansky, A.; Functional analysis, Blaisdell Publ. Co.; New York-Toronto-London (1964).

[60] Wong, Y-C.; Locally o-convex Riesz spaces, Proc. London Math. Soc. (3), 19 (1969), 289-309.

[61] Wong, Y-C.; Order-infrabarrelled Riesz spaces, Math. Ann. 183 (1969), 17-32.

[62] Wong, Y-C.; Open decompositions on ordered convex spaces, Proc. Camb. Phil. Soc. 74 (1973), 49-59.

[63] Wong, Y-C and Ng, K-F.; Partially ordered topological vector spaces, Oxford Math. Monographs (1973).

INDEX

Vol. 521: G. Cherlin, Model Theoretic Algebra - Selected Topics. IV, 234 pages. 1976.

Vol. 522: C. O. Bloom and N. D. Kazarinoff, Short Wave Radiation Problems in Inhomogeneous Media: Asymptotic Solutions. V. 104 pages. 1976.

Vol. 523: S. A. Albeverio and R. J. Høegh-Krohn, Mathematical Theory of Feynman Path Integrals. IV, 139 pages. 1976.

Vol. 524: Séminaire Pierre Lelong (Analyse) Année 1974/75. Edité par P. Lelong. V, 222 pages. 1976.

Vol. 525: Structural Stability, the Theory of Catastrophes, and Applications in the Sciences. Proceedings 1975. Edited by P. Hilton. VI, 408 pages. 1976.

Vol. 526: Probability in Banach Spaces. Proceedings 1975. Edited by A. Beck. VI, 290 pages. 1976.

Vol. 527: M. Denker, Ch. Grillenberger, and K. Sigmund, Ergodic Theory on Compact Spaces. IV, 360 pages. 1976.

Vol. 528: J. E. Humphreys, Ordinary and Modular Representations of Chevalley Groups. III, 127 pages. 1976.

Vol. 529: J. Grandell, Doubly Stochastic Poisson Processes. X, 234 pages. 1976.

Vol. 530: S. S. Gelbart, Weil's Representation and the Spectrum of the Metaplectic Group. VII, 140 pages. 1976.

Vol. 531: Y.-C. Wong, The Topology of Uniform Convergence on Order-Bounded Sets. VI, 163 pages. 1976.

Vol. 532: Théorie Ergodique. Proceedings 1973/1974. Edité par J.-P. Conze and M. S. Keane. VIII, 227 pages. 1976.

Vol. 533: F. R. Cohen, T. J. Lada, and J. P. May, The Homology of Iterated Loop Spaces. IX, 490 pages. 1976.

Vol. 534: C. Preston, Random Fields. V, 200 pages. 1976.

Vol. 535: Singularités d'Applications Differentiables. Plans-sur-Bex. 1975. Edité par O. Burlet et F. Ronga. V, 253 pages. 1976.

Vol. 536: W. M. Schmidt, Equations over Finite Fields. An Elementary Approach. IX, 267 pages. 1976.

Vol. 537: Set Theory and Hierarchy Theory. Bierutowice, Poland 1975. A Memorial Tribute to Andrzej Mostowski. Edited by W. Marek, M. Srebrny and A. Zarach. XIII, 345 pages. 1976.

Vol. 538: G. Fischer, Complex Analytic Geometry. VII, 201 pages. 1976.

Vol. 539: A. Badrikian, J. F. C. Kingman et J. Kuelbs, Ecole d'Eté de Probabilités de Saint Flour V-1975. Edité par P.-L. Hennequin. IX, 314 pages. 1976.

Vol. 540: Categorical Topology, Proceedings 1975. Edited by E. Binz and H. Herrlich. XV, 719 pages. 1976.

Vol. 541: Measure Theory, Oberwolfach 1975. Proceedings. Edited by A. Bellow and D. Kölzow. XIV, 430 pages. 1976.

Vol. 542: D. A. Edwards and H. M. Hastings, Čech and Steenrod Homotopy Theories with Applications to Geometric Topology. VII, 296 pages. 1976.

Vol. 543: Nonlinear Operators and the Calculus of Variations, Bruxelles 1975. Edited by J. P. Gossez, E. J. Lami Dozo, J. Mawhin, and L. Waelbroeck, VII, 237 pages. 1976.

Vol. 544: Robert P. Langlands, On the Functional Equations Satisfied by Eisenstein Series. VII, 337 pages. 1976.

Vol. 545: Noncommutative Ring Theory. Kent State 1975. Edited by J. H. Cozzens and F. L. Sandomierski. V, 212 pages. 1976.

Vol. 546: K. Mahler, Lectures on Transcendental Numbers. Edited and Completed by B. Diviš and W. J. Le Veque. XXI, 254 pages. 1976.

Vol. 547: A. Mukherjea and N. A. Tserpes, Measures on Topological Semigroups: Convolution Products and Random Walks. V, 197 pages. 1976.

Vol. 548: D. A. Hejhal, The Selberg Trace Formula for PSL (2,IR). Volume I. VI, 516 pages. 1976.

Vol. 549: Brauer Groups, Evanston 1975. Proceedings. Edited by D. Zelinsky. V, 187 pages. 1976.

Vol. 550: Proceedings of the Third Japan - USSR Symposium on Probability Theory. Edited by G. Maruyama and J. V. Prokhorov. VI, 722 pages. 1976.

Vol. 551: Algebraic K-Theory, Evanston 1976. Proceedings. Edited by M. R. Stein. XI, 409 pages. 1976.

Vol. 552: C. G. Gibson, K. Wirthmüller, A. A. du Plessis and E. J. N. Looijenga. Topological Stability of Smooth Mappings. V, 155 pages. 1976.

Vol. 553: M. Petrich, Categories of Algebraic Systems. Vector and Projective Spaces, Semigroups, Rings and Lattices. VIII, 217 pages. 1976.

Vol. 554: J. D. H. Smith, Mal'cev Varieties. VIII, 158 pages. 1976.

Vol. 555: M. Ishida, The Genus Fields of Algebraic Number Fields. VII, 116 pages. 1976.

Vol. 556: Approximation Theory. Bonn 1976. Proceedings. Edited by R. Schaback and K. Scherer. VII, 466 pages. 1976.

Vol. 557: W. Iberkleid and T. Petrie, Smooth S^1 Manifolds. III, 163 pages. 1976.

Vol. 558: B. Weisfeiler, On Construction and Identification of Graphs. XIV, 237 pages. 1976.

Vol. 559: J.-P. Caubet, Le Mouvement Brownien Relativiste. IX, 212 pages. 1976.

Vol. 560: Combinatorial Mathematics, IV, Proceedings 1975. Edited by L. R. A. Casse and W. D. Wallis. VII, 249 pages. 1976.

Vol. 561: Function Theoretic Methods for Partial Differential Equations. Darmstadt 1976. Proceedings. Edited by V. E. Meister, N. Weck and W. L. Wendland. XVIII, 520 pages. 1976.

Vol. 562: R. W. Goodman, Nilpotent Lie Groups: Structure and Applications to Analysis. X, 210 pages. 1976.

Vol. 563: Séminaire de Théorie du Potentiel. Paris, No. 2. Proceedings 1975-1976. Edited by F. Hirsch and G. Mokobodzki. VI, 292 pages. 1976.

Vol. 564: Ordinary and Partial Differential Equations, Dundee 1976. Proceedings. Edited by W. N. Everitt and B. D. Sleeman. XVIII, 551 pages. 1976.

Vol. 565: Turbulence and Navier Stokes Equations. Proceedings 1975. Edited by R. Temam. IX, 194 pages. 1976.

Vol. 566: Empirical Distributions and Processes. Oberwolfach 1976. Proceedings. Edited by P. Gaenssler and P. Révész. VII, 146 pages. 1976.

Vol. 567: Séminaire Bourbaki vol. 1975/76. Exposés 471-488. IV, 303 pages. 1977.

Vol. 568: R. E. Gaines and J. L. Mawhin, Coincidence Degree, and Nonlinear Differential Equations. V, 262 pages. 1977.

Vol. 569: Cohomologie Etale SGA 4½. Séminaire de Géométrie Algébrique du Bois-Marie. Edité par P. Deligne. V, 312 pages. 1977.

Vol. 570: Differential Geometrical Methods in Mathematical Physics, Bonn 1975. Proceedings. Edited by K. Bleuler and A. Reetz. VIII, 576 pages. 1977.

Vol. 571: Constructive Theory of Functions of Several Variables, Oberwolfach 1976. Proceedings. Edited by W. Schempp and K. Zeller. VI. 290 pages. 1977

Vol. 572: Sparse Matrix Techniques, Copenhagen 1976. Edited by V. A. Barker. V, 184 pages. 1977.

Vol. 573: Group Theory, Canberra 1975. Proceedings. Edited by R. A. Bryce, J. Cossey and M. F. Newman. VII, 146 pages. 1977.

Vol. 574: J. Moldestad, Computations in Higher Types. IV, 203 pages. 1977.

Vol. 575: K-Theory and Operator Algebras, Athens, Georgia 1975. Edited by B. B. Morrel and I. M. Singer. VI, 191 pages. 1977.

Vol. 576: V. S. Varadarajan, Harmonic Analysis on Real Reductive Groups. VI, 521 pages. 1977.

Vol. 577: J. P. May, E_∞ Ring Spaces and E_∞ Ring Spectra. IV, 268 pages. 1977.

Vol. 578: Séminaire Pierre Lelong (Analyse) Année 1975/76. Edité par P. Lelong. VI, 327 pages. 1977.

Vol. 579: Combinatoire et Représentation du Groupe Symétrique, Strasbourg 1976. Proceedings 1976. Edité par D. Foata. IV, 339 pages. 1977.

Vol. 580: C. Castaing and M. Valadier, Convex Analysis and Measurable Multifunctions. VIII, 278 pages. 1977.

Vol. 581: Séminaire de Probabilités XI, Université de Strasbourg. Proceedings 1975/1976. Edité par C. Dellacherie, P. A. Meyer et M. Weil. VI, 574 pages. 1977.

Vol. 582: J. M. G. Fell, Induced Representations and Banach *-Algebraic Bundles. IV, 349 pages. 1977.

Vol. 583: W. Hirsch, C. C. Pugh and M. Shub, Invariant Manifolds. IV, 149 pages. 1977.

Vol. 584: C. Brezinski, Accélération de la Convergence en Analyse Numérique. IV, 313 pages. 1977.

Vol. 585: T. A. Springer, Invariant Theory. VI, 112 pages. 1977.

Vol. 586: Séminaire d'Algèbre Paul Dubreil, Paris 1975–1976 (29ème Année). Edited by M. P. Malliavin. VI, 188 pages. 1977.

Vol. 587: Non-Commutative Harmonic Analysis. Proceedings 1976. Edited by J. Carmona and M. Vergne. IV, 240 pages. 1977.

Vol. 588: P. Molino, Théorie des G-Structures: Le Problème d'Equivalence. VI, 163 pages. 1977.

Vol. 589: Cohomologie l-adique et Fonctions L. Séminaire de Géométrie Algébrique du Bois-Marie 1965–66, SGA 5. Edité par L. Illusie. XII, 484 pages. 1977.

Vol. 590: H. Matsumoto, Analyse Harmonique dans les Systèmes de Tits Bornologiques de Type Affine. IV, 219 pages. 1977.

Vol. 591: G. A. Anderson, Surgery with Coefficients. VIII, 157 pages. 1977.

Vol. 592: D. Voigt, Induzierte Darstellungen in der Theorie der endlichen, algebraischen Gruppen. V, 413 Seiten. 1977.

Vol. 593: K. Barbey and H. König, Abstract Analytic Function Theory and Hardy Algebras. VIII, 260 pages. 1977.

Vol. 594: Singular Perturbations and Boundary Layer Theory, Lyon 1976. Edited by C. M. Brauner, B. Gay, and J. Mathieu. VIII, 539 pages. 1977.

Vol. 595: W. Hazod, Stetige Faltungshalbgruppen von Wahrscheinlichkeitsmaßen und erzeugende Distributionen. XIII, 157 Seiten. 1977.

Vol. 596: K. Deimling, Ordinary Differential Equations in Banach Spaces. VI, 137 pages. 1977.

Vol. 597: Geometry and Topology, Rio de Janeiro, July 1976. Proceedings. Edited by J. Palis and M. do Carmo. VI, 866 pages. 1977.

Vol. 598: J. Hoffmann-Jørgensen, T. M. Liggett et J. Neveu, Ecole d'Eté de Probabilités de Saint-Flour VI – 1976. Edité par P.-L. Hennequin. XII, 447 pages. 1977.

Vol. 599: Complex Analysis, Kentucky 1976. Proceedings. Edited by J. D. Buckholtz and T. J. Suffridge. X, 159 pages. 1977.

Vol. 600: W. Stoll, Value Distribution on Parabolic Spaces. VIII, 216 pages. 1977.

Vol. 601: Modular Functions of one Variable V, Bonn 1976. Proceedings. Edited by J.-P. Serre and D. B. Zagier. VI, 294 pages. 1977.

Vol. 602: J. P. Brezin, Harmonic Analysis on Compact Solvmanifolds. VIII, 179 pages. 1977.

Vol. 603: B. Moishezon, Complex Surfaces and Connected Sums of Complex Projective Planes. IV, 234 pages. 1977.

Vol. 604: Banach Spaces of Analytic Functions, Kent, Ohio 1976. Proceedings. Edited by J. Baker, C. Cleaver and Joseph Diestel. VI, 141 pages. 1977.

Vol. 605: Sario et al., Classification Theory of Riemannian Manifolds. XX, 498 pages. 1977.

Vol. 606: Mathematical Aspects of Finite Element Methods. Proceedings 1975. Edited by I. Galligani and E. Magenes. VI, 362 pages. 1977.

Vol. 607: M. Métivier, Reelle und Vektorwertige Quasimartingale und die Theorie der Stochastischen Integration. X, 310 Seiten. 1977.

Vol. 608: Bigard et al., Groupes et Anneaux Reticulés. XIV, 334 pages. 1977.

Vol. 609: General Topology and Its Relations to Modern Analysis and Algebra IV. Proceedings 1976. Edited by J. Novák. XVIII, 225 pages. 1977.

Vol. 610: G. Jensen, Higher Order Contact of Submanifolds of Homogeneous Spaces. XII, 154 pages. 1977.

Vol. 611: M. Makkai and G. E. Reyes, First Order Categorical Logic. VIII, 301 pages. 1977.

Vol. 612: E. M. Kleinberg, Infinitary Combinatorics and the Axiom of Determinateness. VIII, 150 pages. 1977.

Vol. 613: E. Behrends et al., L^p-Structure in Real Banach Spaces. X, 108 pages. 1977.

Vol. 614: H. Yanagihara, Theory of Hopf Algebras Attached to Group Schemes. VIII, 308 pages. 1977.

Vol. 615: Turbulence Seminar, Proceedings 1976/77. Edited by P. Bernard and T. Ratiu. VI, 155 pages. 1977.

Vol. 616: Abelian Group Theory, 2nd New Mexico State University Conference, 1976. Proceedings. Edited by D. Arnold, R. Hunter and E. Walker. X, 423 pages. 1977.

Vol. 617: K. J. Devlin, The Axiom of Constructibility: A Guide for the Mathematician. VIII, 96 pages. 1977.

Vol. 618: I. I. Hirschman, Jr. and D. E. Hughes, Extreme Eigen Values of Toeplitz Operators. VI, 145 pages. 1977.

Vol. 619: Set Theory and Hierarchy Theory V, Bierutowice 1976. Edited by A. Lachlan, M. Srebrny, and A. Zarach. VIII, 358 pages. 1977.

Vol. 620: H. Popp, Moduli Theory and Classification Theory of Algebraic Varieties. VIII, 189 pages. 1977.

Vol. 621: Kauffman et al., The Deficiency Index Problem. VI, 112 pages. 1977.

Vol. 622: Combinatorial Mathematics V, Melbourne 1976. Proceedings. Edited by C. Little. VIII, 213 pages. 1977.

Vol. 623: I. Erdelyi and R. Lange, Spectral Decompositions on Banach Spaces. VIII, 122 pages. 1977.

Vol. 624: Y. Guivarc'h et al., Marches Aléatoires sur les Groupes de Lie. VIII, 292 pages. 1977.

Vol. 625: J. P. Alexander et al., Odd Order Group Actions and Witt Classification of Innerproducts. IV, 202 pages. 1977.

Vol. 626: Number Theory Day, New York 1976. Proceedings. Edited by M. B. Nathanson. VI, 241 pages. 1977.

Vol. 627: Modular Functions of One Variable VI, Bonn 1976. Proceedings. Edited by J.-P. Serre and D. B. Zagier. VI, 339 pages. 1977.

Vol. 628: H. J. Baues, Obstruction Theory on the Homotopy Classification of Maps. XII, 387 pages. 1977.

Vol. 629: W. A. Coppel, Dichotomies in Stability Theory. VI, 98 pages. 1978.

Vol. 630: Numerical Analysis, Proceedings, Biennial Conference, Dundee 1977. Edited by G. A. Watson. XII, 199 pages. 1978.

Vol. 631: Numerical Treatment of Differential Equations. Proceedings 1976. Edited by R. Bulirsch, R. D. Grigorieff, and J. Schröder. X, 219 pages. 1978.

Vol. 632: J.-F. Boutot, Schéma de Picard Local. X, 165 pages. 1978.

Vol. 633: N. R. Coleff and M. E. Herrera, Les Courants Résiduels Associés à une Forme Méromorphe. X, 211 pages. 1978.

Vol. 634: H. Kurke et al., Die Approximationseigenschaft lokaler Ringe. IV, 204 Seiten. 1978.

Vol. 635: T. Y. Lam, Serre's Conjecture. XVI, 227 pages. 1978.

Vol. 636: Journées de Statistique des Processus Stochastiques, Grenoble 1977, Proceedings. Edité par Didier Dacunha-Castelle et Bernard Van Cutsem. VII, 202 pages. 1978.

Vol. 637: W. B. Jurkat, Meromorphe Differentialgleichungen. VII, 194 Seiten. 1978.

Vol. 638: P. Shanahan, The Atiyah-Singer Index Theorem, An Introduction. V, 224 pages. 1978.

Vol. 639: N. Adasch et al., Topological Vector Spaces. V, 125 pages. 1978.